This book is dedicated to my parents

Joseph Andrew Wolf

and

Dorothy Jane Wolf

Contents

Acknowledgements

This book began as my doctoral dissertation in the Critical Studies Department in the School of Cinema/Television at the University of Southern California, and I would like to thank my dissertation committee, Marsha Kinder, Michael Renov, and Richard Weinberg, for their help and encouragement, and the generous support they gave me in my years at USC. Their enthusiasm kept me going, and I am very grateful for their help. I would also like to thank Gene Coe, A. Michael Noll, Michael Heim, and Todd Drow, for reading various drafts of chapters, while it was still a dissertation. I am am also grateful to those who used this work in other venues; part of chapter 4 appeared as "In the Frame of *Roger Rabbit*: Visual Compositing in Film" in *The Velvet Light Trap*, No. 36, Fall 1995, and part of chapter ten appeared as the essay "Subjunctive Documentary: Computer Imaging and Simulation", in *Collecting Visible Evidence*, an anthology edited by Michael Renov and Jane Gaines and published in 1999. And, of course, a big thanks goes to my parents, to whom I dedicate this book, and thanks to God who made it all possible.

PREFACE

Digital technology is among the fastest-growing phenomena of recent decades, with the appearance of affordable computers and their integration into a wide range of technologies, most notably communications technology. The shift from analog technology to digital technology, however, has repercussions often left unacknowledged or unexplored in depth. These differences stem not only from the applications of the technology but their basic, inherent nature as well. A good place to begin, then, might be the differences between "digital" and "analog", and the concepts they represent.

While the concept of "digital" today often pertains to intangible abstractions, its origins lie in the realm of the tactile. "Digit" comes from the Latin *digitus*, meaning a finger or a toe, a definition which the *Oxford English Dictionary* tells us is "Now only *humorous* or *affected*."[1] Today, of course, the term refers mainly to the numerals zero through nine, the basic elements of the number system. The fourth and fifth definitions given by the *OED*, however, are relatively recent additions relating to computer terminology:

> **4.** Of, pertaining to, or using digits [DIGIT *sb.* 3]; *spec.* applied to a computer which operates on data in the form of digits or similar discrete elements (opp. *analogue computer*).
> **5. a.** Designating (a) recording in which the original waveform is digitally encoded and the information in it represented by the presence or absence of pulses of equal strength, making it less subject to degradation than a conventional analogue signal; of or pertaining to such a recording.[2]

In the past few decades, these new definitions have made the term "digital" almost synonymous with the computer, despite a long history of digital technologies preceding the computer.

The term "digital" has a number of other connotations. Conceptually, digitization is often connected with *quantization*, a process closely related but not synonymous to it. While digitization concerns the conversion of data into numeric form, to *quantize* something is to restrict the values or states of a system so that variables can only appear at discrete magnitudes which are multiples of a common unit. In other words, quantization sets a number of distinct levels or units which are used to measure something, and these are the *only* levels at which data can be represented. For example, when a student receives a grade in a class, the grade will be either A, A-, B+, B, B-, C+, C, C-, and so on; you can't get something in between (at least at most schools). A student's performance, then, is rounded off to the nearest grade level, to make the whole process of grading easier and simpler. The same thing happens when sound is quantized; the continuous sound wave is broken up into "samples", each of which is rounded off to one of the possible levels allowed by the machine which is quantizing it. Likewise, pictures are broken up into *pixels*, the tiny square elements that make up computer imagery. Quantizing, then, takes something analog with infinite detail or gradations, and simplifies it into something with a limited amount of detail, making it easier to work with or "store" as data in a limited amount of computer memory.

Once the data is broken up into pieces (like grades, samples, pixels, etc.), those pieces can be represented by numbers and encoded into numerical form; this is the basis of *digitizing*.[3] Thus, in an analog-to-digital conversion, some form of quantizing must occur before digitizing can occur. When "digitizing" is referred to as a process, it more often than not includes quantizing along with it, since the two processes are so closely related; but it still is important to see two processes as distinct, one preceding the other. Likewise, a societal trend towards digitization can only occur after a quantization of everyday life occurs —a process which is the subject of chapter one.

The effects of quantization and digitization are perhaps best described in two keys words, *discrete* and *representation*. The discrete nature of digital data is what separates digital and analog forms, and encoding changes the form of representation. This change is illustrated by the difference between analog and digital computers: analog computers use physically measurable quantities (length, weight, voltage, etc.) to represent numbers, while digital computers use symbolic representations of variables. Thus the conversion from analog to digital involves a semiotic shift from the indexical to the symbolic, a topic which will be examined in detail in the final chapter on indexicality.

Another difference between analog and digital is their connotative meaning in popular usage. In common parlance, "digital" has come to represent the modern, state-of-the-art technology, while "analog" refers to an older, outmoded and outdated form; this distinction is perhaps most obvious in the music industry where compact discs quickly replaced vinyl albums as the dominant commercial format. The term "digital" is also often associated with a high degree of quality, even though the term refers to a technology and not a specific application of it. What is usually not acknowledged is that all output devices, monitors, speakers, printers, and so on, are analog devices. To be of use, sound and image must reenter the domain of the physical world, and in doing so there is an inevitable shift back to analog form. This is similar to the idea that no one has ever seen a perfect circle. A circle is only perfect when it exists as a mathematical entity; once it is drawn up or printed out, imperfections in physical media, albeit small ones, render it imperfect. Thus "digital" can only refer to data represented or "stored" in digital form; as output it becomes analog again.

It would seem, then, that "digital image" is oxymoronic; if stored in numeric form, the image is not an image in the conventional physical sense; we cannot see it. And once it *is* in visible form, as output, it is no longer strictly digital. The term "digital image" does make sense if we refer to another definition of "image", that of a mental picture or representation of something. Thus, when "digital image" is used here, it will mean an image which has been stored in some digital format. In this sense, the digital image promotes a shift from the *perceptual* to the *conceptual*, a theme running throughout this book. With the widespread and full-scale integration of digital technology into daily life, "digital" has become more than a type of technology; it has come to stand for the fabric of the growing information society. It extends beyond a form of design into a way of thinking, an attitude towards the world and the future.

The first part of this book, *The Emergence of Digital Technology*, examines the frame of mind and ways of thinking from which digital technology arose, and the conditions which made it desirable. Chapter one looks at the quantization of everyday life that set the stage for digital technology, and chapter two looks at the development of digital technology and a variety of its precursors.

The second part, *Art*, is concerned with the effects of digital technology within art and culture. Chapter three looks at how digital technology has been integrated into the production, preservation, exhibition, and reproduction of art, and changes in the notion of "art" itself, and it explores the implications of digital artwork's lack of physicality, tracing its links to the physical world. Chapter four looks at how the technological basis of digital artwork results in biases which can occur at the cultural level, and chapter five looks at how digital technology has expanded the possibilities of composite imagery, and some of the implications of these changes.

The third part, *Communication / Media*, builds on second part and broadens its scope out from art to include all other forms of communication and media, and looks at how digital technology has been positioned among them. Chapter six examines the growth of machine mediation in social interaction and notions of interactivity, while chapter seven looks at the culmination of electronic communication with the forging of the conceptual, informational realm known as cyberspace.

The final part, *Perception / Representation / Cognition*, extends the scope of previous chapters to include the activities through which individuals perceive and understand the world around them. Chapter eight examines the effects of digital technology on the environment, and the way it recreates the user's environment. Chapter nine continues this theme and looks at notions of virtual reality, the fantasies surrounding them, and the idea of the substitute. Finally, chapter ten, on indexicality, traces how digital technology mediates and abstracts the indexical linkages between the observer and the observed, and how the notion of indexicality itself has been called into question. It looks at the implications of these changes for users, and the increasing degrees of abstraction brought about by digital technology and the media.

Throughout this book, I try to show how the effects of digital technology, and the concepts surrounding it, have subtly altered the fabric of society and culture in areas both the practical and theoretical. While many of these changes may appear, initially, to be small, insignificant, or even for the better, it is not so easy to determine the worth of their combined effects, arising from the implementation of the technology and the nature of the technology itself. It is my hope that this book will help to open up inquiry into these areas.

NOTES

1. *The Oxford English Dictionary, Second Edition, Volume IV: Creel-Duzepere*, prepared by John. A. Simpson and Edmund. S. C. Weiner, New York: Claredon Press, and Oxford, England: Oxford University Press, London, ©1993, page 653.
2. Ibid., page 654.
3. Occasionally, because "digital" is the adjective form of "digit", people mistakenly use "digitalize" in place of "digitize"; *digitalization* refers to the administration of medicine prepared from *digitalis*, a genus of plant including the foxglove, or *Fingerhut*, the German name for the plant.

I.

Digital Development

1.
The Quantization of Everyday Life

Consider how many numbers you use or see on a daily basis, and the important role they play in giving order to your life and the way in which you picture and think about the world. Even such activities as talking on the telephone or buying something at a supermarket involve digital technologies and the streams of numbers they produce. In the last century or so, numbers have taken on an important and often central role in people's lives, allowing digital technologies to flourish. Of course these technologies are partly responsible for the increase as well; but they could only come about in a world already obsessed with measuring and counting.

The concept of "digital", and technology based on it, required the idea of counting, the notion of *quantity*, and the ability to see things as distinct entities or separable into distinct parts. Being able to think of things as made up of component parts is immensely useful as a way of thinking, leading to new ways of seeing the world. But in recent times it has been taken to quite an extreme —and is encouraged by modern digital technology itself.

Digital technology promotes a quantized style of thinking that produces a limited, if not hazardous, way of looking at the world, changing the nature of cognition and the individual's link to lived experience and intersubjective reality. (When I use the word "quantizing" here, I mean it in a conceptual sense; how we think of things, not necessarily the things themselves; the *signifier*, not the *signified.*)

While widespread promotion and acceptance of a "quantized" way of thinking has been relatively recent, its roots precede the digital age, extending back through recorded history. Although quantization certainly is useful and has been essential in shaping much of Western culture (and to a degree Eastern culture), it also has limitations and disturbing side effects.

Divide and Conquer

In order to make sense of the world —or rather, our sensory impressions of it— we break it up perceptually into series of parts; through visual cues like color, focusing depth, motion, shape and texture, we distinguish individual objects in the visual field, separating foreground from background. In a similar fashion, we also break things up conceptually, in order to name them and refer to them. The color spectrum is continuous, yet it has been divided up into colors which are named; although people may not always agree on the boundaries (between 'red' and 'orange', for example) references to specific colors are generally understood.

Such boundaries are artificial ones, but are important in defining the objects of study in question. Historians are aware of the perils of periodization, which imposes an order the structure of which is determined by certain events at the expense of others. We might think of the 1940s, 1950s or 1960s as separate, distinct periods, even though more cumbersome divisions of 1951-1957, 1958-1962, and 1963-1971 might prove just as useful historically. The division by decade is merely a numerical one, with no reference whatsoever to the history being periodized.

Not only are things divided, but divisions tend to be of equal size or measure, standardized so that consistent and interchangeable units are created; unit multiples can be calculated quickly, and measurements made by one person will be consistent with those made by another. This is similar to the idea behind *quantization*, the process in which an analog range of values is made to fit into a finite number of discrete levels or units, usually equal in size, so as to be represented more simply. This simplification made understanding, representing, and remembering easier and communication more precise. Quantization, as a form of 'rounding off', is simplification at the expense of accuracy, and throughout history, as we shall see, attempts to regain accuracy have been made through the use of increasingly smaller units.

Quantization, then, arose out of the 'divide and conquer' thinking that successfully had allowed people to break down and reconstruct the world conceptually, and communicate ideas about the world and the objects in it. Archaeologists suggest that written communication itself may have arisen from mathematical representation when thousands of years ago, tally marks used for measuring amounts grew into more expressive forms.[1] The invention of mathematics was the first step towards quantization, because it allowed *quantification*, the expression of things as quantities, in amounts or numbers. Since many things being measured were not conveniently or consistently divided by nature into individual objects, arbitrary units and measures came into being which were first based on nature but which gradually tended towards complete abstraction. The contents of everyday life were broken up and rounded off into these units, to make life in general a more orderly experience.

The quantization of everyday life can be seen in four conceptual areas which have been broken up into units and continuously subdivided and abstracted into increasingly smaller units for greater manipulation and interchangeability. These concepts, *Time*, *Space*, *Value*, and *Information*, are four constructs we use to mentally reconstruct and order the way we think of the world; changes in the way we conceptualize them become changes in cognition itself.

Time

Without reliably regular intervals produced by natural events such as sunrise and sunset or the phases of the moon, how could one consistently measure the passage of time? What can be relied upon to measure time, apart from consistencies found in nature? And how can we be sure that they really are consistent, if they are the basis of the measuring devices themselves? These are some of the questions that had to be overcome in time measurement, and measures of space relying on time measurement.

Timekeeping began at the dawn of history, and the desire for consistency and precision has driven timekeeping developments ever since. Technological advancements have allowed for increasingly finer units of time to be measured, and have worked in tandem with people's desire or need to keep track of smaller and smaller units. As A. J. Turner writes in *Of Time and Measurement: Studies in the History of Horology and Fine Technology*;

> ... smaller units had to be imagined, and ways to determine them devised. Such active time-measurement involved the development of tools and machines. It also involved a change of relationship with time. Gradually time became more manipulable. Time as a given element of the world gave way to a time which was created by the machines which measured it. In the process man became increasingly independent of nature. Whether this be seen as a liberation or denaturisation, it was an important consequence of the development of time-measuring devices and is apparently irreversible.[2]

The earliest measurements of time arose from the observation of cyclical or regular phenomena in Nature. Sunrise, sunset, full moon, new moon, and the positions of constellations were based on celestial mechanics, while the flooding of the Nile helped the ancient Egyptians determine the length of the year. The combination of the two natural units of time, the day and the year, resulted in the calendar, which the Egyptians set at 365 days, possibly as early as 4228 BC.[3]

The next division of time was the marking of noon —the sun's zenith— and the division of the day into hours. Although the earliest evidence of sundials dates from around 2000 BC, divisions of the day into hours of equal length were first used by astronomers and the physicians and astrologers who relied on the astronomer's work, and it was around the 14th century that hours finally became more commonly used in social life. Turner points out that there was even some resistance to hourly divisions, due to the

> ...clash between biological time and the artificial time of the sundial when this was adopted in social life. That the tiresomeness of waiting for the sundial to give one leave to eat furnished matter for the comic poets Plautus (3rd century BC) and Alciphron (1st century AD) is suggestive of a more widely spread reaction.[4]

Eventually the division into hours became accepted, especially in regulated communities like Christian monasteries, and new technologies such as the weight-driven clock and later the pendulum clock lent timekeeping greater precision. During this time of public acceptance, craftsmen made clockmaking into an art, and the clock took on greater importance in daily life, and even a town's self image.[5]

During the Renaissance, clock movements became smaller, and the clock moved indoors and entered into family life. In the early 16th century, further miniaturization brought about the carriage-clock, and finally the pocket-watch. Mechanical accuracy and the recognition of the importance of time had increased to the point where minute hands came into use and were seen as necessary.[6] The next division of time divided minutes into seconds. Although clocks with a second hand appeared during the 16th century, their accuracy was still far from being good enough to warrant having one, nor was there any societal need for one. Even in contemporary life, there is scarcely little need for a second hand, except to indicate that the clock or watch is still running and has not stopped; most digital light-emitting diode clocks do not even display seconds. Why then, were timepieces with second hands produced in an era that did not have the technology for the needed precision? Historian Carlo M. Cipolla suggests an answer;

> The most striking occurrence in the early history of clocks is that while medieval craftsmen did not improve noticeably in precision, they soon succeeded in constructing clocks with curious and very complicated movements. It was easier to add wheels to wheels than to find better ways to regulate the escapement. On the other hand complicated movements had quite a popular appeal and most people believed that a correct knowledge of the conjunction of the heavenly bodies was essential for the success of human enterprises.[7]

Already, technology was beginning to outstrip the public's ability to understand it. It had become something of a novelty, and an aid to growing astrological superstitions.

By 1896, the year of the return of the Olympic Games, Olympic records were being recorded with an accuracy that included tenths of a second, and by 1968 they were recorded in hundredths of a second. Science, too, had grown in complexity, requiring ever finer measurements of time, space, and mass. In the late 1920s, Joseph W. Horton and Warren A. Marrison of Bell Labs attained a new level of accuracy with the first quartz-crystal clock, which became the primary laboratory standard by the 1940s. But physics demanded even greater precision, for example, in experiments testing for relativistic time dilation, in which a clock aboard an airplane was predicted to run billionths of a second faster than one on the ground. Atomic clocks, based on the oscillations of atoms, provided the answer. In 1967, the second was redefined atomically as being equal to 9,192,631,770 oscillations of the cesium-133 atom.[8]

Just as the calendar regulated life, advances in timekeeping precision led to adjustments made to the calendar. The year is slightly longer than the 365 days the Egyptians measured, and Hellenistic astronomers added a leap day to make up for the missing quarter day every year. The leap day was officially adopted into the calendar in 46 BC, in Rome under the reign of Julius Caesar. In 1582, Pope Gregory XIII's advisors persuaded him to drop the leap day in years ending with two zeroes, since the year was not quite 365.25 days either. And more recently, in 1987 and 1992, timekeepers have added "leap seconds" to restore accuracy; the earth's spin on its axis has been slowing down by about one millisecond (a thousandth of a second) per day, and leap seconds help the earth to catch up to humanity's clocks.[9]

But the quantization of time does not end there; theoretical physics went further still, speculating on what the shortest spans of time possible might be. One proposal was the *chronon*, described as the time taken for a photon to traverse the diameter of an electron, which is approximately equal to 10^{-24} seconds. And finally, the smallest unit of time ever conceived is the unit known as *Planck-time*, which is the amount of time it takes for a photon to move through a distance equal to one unit of *Planck-length* (equal to 10^{-35} meters). One unit of Planck-time is equal to 10^{-43} seconds, an unimaginably small amount of time; there are more units of Planck-time in one second than there are seconds in the current age of the universe —more, in fact, than if the universe was 21,125,500 *billion billion* times the age it is now!

Clearly time, our experience of it, and its value to us, have changed the way we think of the world. Time has ceased to be a continuous flow, and become fragmented and segmented; most public events in daily life are set to begin on the hour or half-hour, and private ones often are as well. We think of time as having arbitrary units of equal length, subdividing them and grouping them into larger units as well. Conceptually, our temporal life has become discontinuous and departmentalized; the irony is that the more divisions we make in our day, the less time we seem to have.

Space

The quantization of time has always been closely related with the measurement of distance; in early societies day and night and the phases of the moon may have been the only ways of expressing distances. Astrolabes, invented during the Middle Ages, provided a means of measuring large distances or heights using celestial bodies as a guide. The sun, moon, and stars were used for navigation and aided cartographers in mapping land and sea; they even allowed Eratosthenes, a Greek astronomer, to estimate the circumference of the earth in 200 BC. Measurements of smaller distances were also based on nature, and often on the human body, which was always available for use and easy to understand. But the disadvantages of such a system were the lack of common multiples or divisions of units, as well as the varying size of the body from one person to another.

Like time measurement, the need for standards took linear measurement into abstraction and the devising of more arbitrary units. At first, attempts were made to keep the older system; around the beginning of the 12th century, the yard was set as the distance between King Henry I's nose and the middle fingertip of his outstretched arm. This brought some interchangeability between units, albeit a complicated and convoluted one;

> In England the digit-- later standardized at 3/4 inch (1.905 cm) --was originally a finger's breadth, equal to 1/4 palm, 1/12 span, 1/16 foot, 1/24 cubit, 1/40 step, and 1/80 pace. The palm or hand's breadth was equal to 1/3 span or 1/6 cubit. Based on the foot of 12 inches, it was made equal to 3 inches (7.62 cm). A span was equal to the distance from the tip of the thumb on the outstretched hand, and based on the foot it was made equal to 9 inches (2.286 dm). The cubit was the distance from the elbow to the extremity of the middle finger, which was generally reckoned as 18 inches (4.572 dm), or 6 palms or 2 spans. A step was 1/2 pace or approximately 2 1/2 feet (ca. o.76 m), while a pace equaled 2 steps or approximately 5 feet (ca. 1.52 m). Other body measurements

> were the shaftment of 6 inches (ca. 15.24 cm) or the distance from the tip of the extended thumb across the breadth of the palm; the nail, used principally for cloth, that represented the last two joints of the middle finger, equal to 1/2 finger, 1/4 span, and 1/8 cubit, and standardized at 2 1/4 inches (5.715 cm); the hand of 4 inches (10.16 cm); the finger for cloth equal to 2 nails or 1/2 span, and generally expressed as 4 1/2 inches (1.143 dm), and the fathom, the length of a man's outstretched arms containing generally 6 feet (1.829 m).[10]

In Ronald Edward Zupko's book, *Revolution in Measurement: Western European Weights and Measures Since the Age of Science*, from which the above quote is taken, he recounts the history of British measurement systems and reforms, the coming of the first metric system in France in 1795 and its rapid international spread afterwards, and the reluctant acceptance of the metric system in Britain and the United States after great resistance. The spread of the metric system and its victory over systems in use for centuries was due to its usefulness in science. Metric units are all base 10 and easily convertible, and distance, volume, and weight are all interrelated.

The basic unit of the metric system, the meter, was originally intended to represent one ten-millionth of the distance along the meridian running from the North pole to the equator through Dunkirk, France and Barcelona, Spain. Like many other arbitrarily set standards of measurement, the definitive meter was a metal bar kept as a physical replica. Such physical standards were in the safekeeping of the authorities, but still there was a need for an abstract or mathematical way of defining the standard which could not be destroyed, and which could be calculated anywhere without having to depend on a physical replica in a vault somewhere. Nor was the replica exactly the right length; increased precision measurement of the distance between pole and equator showed the error of the original surveyors to be off by about two miles.[11]

Over the years, technological advancements have continued to redefined the meter; "Up until 1893, the meter was defined as 1,650,764.73 wavelengths in vacuum of the orange-red line of the spectrum of krypton-86. Since then, it is equal to the distance traveled by light in a vacuum in 1/299,792,458 of a second."[12] As was the case with time measurement, science required increasingly finer units of measurement, and the metric system provided with a series of prefixes (1 decimeter= 10^{-1} m; 1 centimeter= 10^{-2} m; 1 millimeter= 10^{-3} m; 1 micrometer= 10^{-6} m; 1 nanometer= 10^{-9} m; 1 picometer= 10^{-12} m; 1 femtometer= 10^{-15} m; 1 attometer= 10^{-18} m). Finally, the smallest unit of length is that of *Planck-length* in theoretical physics, on which the unit of Planck-time is based. One unit of Planck-length is defined as

"The length scale at which a classical description of gravity ceases to be valid, and quantum mechanics must be taken into account... The value of Planck-length is of order 10^{-35} m (twenty orders of magnitude smaller than the size of a proton, 10^{-15} m)."[13] Such a unit is unimaginably small; to compare one unit of Planck-length to the thickness of a sheet of paper would be like comparing the thickness of a sheet of paper to a distance wider than the known universe. Likewise, other distances devised by astronomers are unimaginably large; one AU, or Astronomical Unit, is about 93 million miles, the distance from the earth to the sun; a light year, the distance that light travels in one year, is about 5880 billion miles, and a parsec (from *parallax second*) is 3.26 light years.

In the modern world, numerous factors have changed our sense of space and made us more conscious of its organization. Technological miniaturization, of everything from engines to microchips to household appliances, has changed the nature of the space around us and our relation to it. Microscopes and microphotography have shown us how enormous activity and complex structures can occur in a tiny space, and electrical and molecular engineering have shown how machines can be built at this scale. On the other hand, transportation and communication technologies have shrunk large distances and allowed us to form better and more detailed cognitive maps of the world; although they can sometimes be as distorted as those of ancient cartographers whose biases are often apparent in their maps (the Mercator projection, for example, emphasizes the north and de-emphasizes the south, and Germany, Mercator's homeland, is the projection's centerpoint). The way in which the land and space we occupy are divided can account for much of our experience of that space.

The earliest evidence we have of land division or landscaping is the remaining works and plans of the ancient Egyptians, who were fond of straight-ahead linear arrangements and bilateral symmetry. Straight lines and right angles are a natural product of surveying techniques, which measures distance in straight lines and area as the product of two lengths perpendicular to each other. Egyptian surveying was remarkably precise; the great Pyramid of Giza, for example, is extremely accurate in its dimensions and layout despite its enormous size. The ancient Greeks took the idea of landscape architecture further still, providing the foundation for city planning and land division in Western civilization. Their ancient cities also used the gridiron plan, in which rows of streets are laid out perpendicular to each other in a checkerboard pattern, indifferent to the shape of the land, shorelines, and changes in terrain.[14]

From ancient times onward, the grid pattern, breaking up space in units of equal size and shape, has been forcibly applied onto the land, and onto the surface of the earth in general. Latitude and longitude were introduced by the Greeks around 500 BC, and Eratosthenes, who estimated the earth's circumference, also devised a world map with lines of latitude and longitude, although the lines on these early maps were not evenly spaced; they were drawn to connect places that had the same length of daylight on the longest day of the year. The first uniform grid of parallels and meridians was developed in the second century BC, and is credited to the astronomer Hipparchus. From the stereographic projection of 130 BC and on through to the modern-day Peters projection and satellite mapping, cartographers have tried their hand at squaring the sphere, in their attempts to apply grids and develop flat projections of the earth's surface. In both the gridiron method and the system of longitude and latitude, mapmakers struggle to impose designs onto nature, even when the fit is a forced one due to land formations, terrain, or the curvature of the earth. Because of the earth's curvature, one degree of latitude can vary from 68.703 miles near the equator to 69.407 miles near the poles. Meridians of longitude converge at the poles, so one degree will vary, changing in length from 69.172 miles at the equator to 0 at the poles. The need for quantization again takes measurement and division away from the natural and into the abstract. And the units grow smaller; now that we live in a world where license plates on cars can be seen and read by satellites, degrees of latitude and longitude can be expressed in minutes and seconds of arc, amounting in global coordinate units that are less than a hundred feet wide.[15]

One would think that meridians would provide a convenient way to determine a time zone system, each zone covering an average of 15 degrees of longitude, but it is not as simple as that. Time zones were made necessary by high-speed travel and instantaneous communication during the era of the railroad and the telegraph. Each railroad developed its own time zones, and by the 1870s, 50 different ones were in use. An international conference to establish world time zones was held in 1884 in Washington D. C., but it was not until 1918 that the actual boundaries between the zones were established. Here again, irregularities of nature kept the boundary lines from being straight, and political divisions of land also determined where the boundaries should fall. As a result, there are areas on the globe (in the islands of the Pacific, for example) where, just by traveling north or south, one can change time zones by as much as three hours.

Divisions imposed by meridians and parallels can affect nations, particularly in times of war when territory is in question; for instance, the Mason-Dixon line, set at a latitude of 39° 43' 19.11", divided North and South, separating slave and free states; and the 38th parallel became the line separating North and South Korea during the Korean war. On a smaller scale, lines of longitude and latitude were used to determine much of the town and city planning of early America. On May 20, 1785, the United States Congress authorized the surveying of the western territories into six-mile-square townships, determined by lines of longitude and latitude. Each township was further subdivided into 36 square sections of 640 acres each. Around two-thirds of the present United States were sectioned off in this manner.[16]

The gridiron system was convenient for surveying and land speculation, as every section could be located by number, and deeds were often purchased and recorded that way. But the gridiron plan was not without its detractors, and was sometimes impractical due to the rigidity of the squaring of the land. As John Stilgoe notes in *Common Landscape of America, 1580 to 1845*;

> Roads followed section lines and section lines followed the compass. Surveyors gave no thought to avoiding natural obstacles or approaching natural resources, and as more than one anti-grid congressman had argued in the 1784 and 1796 debates, many settlers suffered permanently. Roads led deliberately and directly through swamps and over hilltops, tiring horses and infuriating drivers. Farmers discovered that some sections were well watered and that others were separated from useful ponds and springs by only several yards. Had the surveyors been allowed to modify the straight lines--even slightly--many sections would have been far more valuable.[17]

As it turned out, the straight lines had to be altered anyway. The use of meridians posed a further problem, because unlike parallels, they converged at the poles, and the distances between them changed about sixty yards per mile. According to Stilgoe,

> What evolved was the section correction, a common design solution to a most vexing geometrical problem. Every few score miles, surveyors shifted the meridian lines a hundred yards farther west and continuing platting. Hardly anyone on the ground noticed the irregularities scarcely visible as two right-angle turns separated by perhaps 300 or 400 yards, and readers of maps found them scarcely more obvious. . . . For all its shortcomings, the grid proved reasonably effective in ordering the land for sale and settlement. People grew accustomed to it, so accustomed in fact that had even the federal government wished to alter it or discard it for some better form, public opposition would have proved too strong. Phrases such as "a square deal" and "he's a four-square man" entered the national vocabulary as expressions of righteousness and fairness. By the 1860s the grid objectified national,

not regional order, and no one wondered at rural space marked by urban rectilinearity.[18]

Land division grids still persist in the United States, in larger and larger scales, and with new construction technologies the land can be made to conform to it more than ever before. Because the grid design is used at so many different scales —globally, regionally, locally, and even at smaller scales like parking lots and tiled plazas— there is often a certain homogeneity to the look of things on vastly different scales. In the film *Koyaanisqatsi* (1983), extreme closeup shots of microchips are intercut with overhead satellite photos of cities at night, the colorful networks eerily resembling one another.

Within the land, there are further attempts at uniformity, as rows of identical highrises and places like Levittown can attest. Vertical space is also quantized, from the earliest terraced cultivation that turned a smooth slope into a series of steps, to the present day when even the airspace above buildings in downtown areas can be bought and sold. Most buildings are divided into numbered floors, and in urban residence towers, it is common for apartments to go up in price as they go up in floor.

In all of these cases, there is an increasing use of numbers as coordinates to locate a person within the grid of the city. Many U.S. cities have series of numbered streets or avenues (or both, like New York City), which are usually an index of a location's distance from City Hall or some natural boundary like a lake. Numbered streets also give some sense of the distance from one point to another; even if we don't know the city, we know that 9th Street is probably two miles away from 34th Street. On the streets themselves, there are house numbers, and even within a residence, an address will often have numbers to further specify a floor, suite, or apartment. As if that is not enough, there is the five-digit zip code, which in 1981 was expanded by an extra four digits tagged on after a hyphen, enabling automated equipment to sort mail down to a specific carrier, the person who makes the delivery.[19] Phone numbers likewise contain an area code, the prefix indicating a neighborhood area, and a country code if called internationally. Oddly enough, electronic mail addresses seem to be the only ones that occasionally get away with having all letters and no numbers in them.

The use of numbers for identification, location, and amount leads us into the areas of information and exchange value, two closely connected areas whose quantization has been slower than those of space and time, partly because what they measure is more abstracted from nature, more culturally variable, always in flux, and more difficult to standardize.

Value

What is value? It is not an inalienable or intrinsic property of matter, so how can it be quantified? Like space and time, the need to measure it exists only when there is a need to communicate it to someone else. Although value itself is always in flux, just as the values of currencies fluctuate, it is still quantized into numerically expressible amounts. The need to quantify value came about with trade and the need to insure that an equitable exchange had been made, since trade was an impetus for the development of mathematics itself. According to financial historian Ray B. Westerfield, the barter system, probably the first system of exchange, was inefficient;

> The dependence on chance coincidence makes barter an inadequate means of developing a market in which anyone can offer his goods with reasonable assurance of being able to trade for something else at least equal in utility to him. ... Occasionally he was forced, in order to make any trade at all, to accept some goods which he did not want for itself but which he knew someone else would be willing to accept in trade for something else he did want. A series of such three-party trades might well establish in each community the habit of looking on some particular goods as widely enough acceptable to act as a satisfactory medium for execution of any exchange. The money idea may also have taken root in a slightly different way. A particular goods may have become generally acceptable due to its basic value, not with the idea of receiving it and holding it between trades, but merely as a common denominator or standard against which to measure the value of both large (or intensely desired) things and small (or only slightly desired) things in working out the terms of a trade. . . . Traces of economic activity in the very earliest civilizations almost invariably show some commodity --cattle, grain, shells, trinkets and the like-- used as an exchange medium. With the passage of the centuries precious metals gained almost complete ascendancy over other commodities as a medium of exchange because they combined the attributes of portability, divisibility, durability, homogeneity, recognizability, and stability of value.[20]

Before value could be quantized, it had to be quantified, and precious metals had the attributes needed for a standard with consistency. Perhaps more than any other precious metal, gold has always been considered valuable, as far back at least as the ancient Egyptians and the Israelites of the Old Testament. Over time, most nations converted over to the gold standard, and in 1900, the United States passed the Currency Act, establishing gold as the currency standard. In *Capital*,

Karl Marx wrote of how one commodity (in this case, gold) becomes the socially-accepted *general equivalent* used to measure the value of all other commodities. The general equivalent, then, simplifies quality into a question of quantity; one bar of gold is as good as any other, as long as it contains the same *amount* of gold.

While measures of weight like the ancient system of shekels, minas, and talents (there were 60 shekels to a mina, and 60 minas to a talent; the shekel was 0.497 ounces or 14.1 grams) were used to regulate amount, a more practical means of quantizing value came around 770-670 BC, when the coin, as a regular and standardized amount of a precious metal, came into being.[21] Each coin was considered equal to every coin of its kind, and all were stamped by the government who vouched for them. All values encountered in trade were expressed in multiples of the coin of lowest denomination; these then, were the discrete levels at which value could appear. Currently in the United States, the smallest unit of value would be the penny, or one-cent piece (although prices at gas stations are still measured out to the mill, or one-tenth of a cent, and then rounded off).

Unlike the units of space and time, the value of any currency or metal standard varies with the economy it exists in. This variation was also aided by the further abstraction of value from gold to less precious metals and to paper money, which first appeared in China around the 9th century AD. The abstraction of money into a more easily reproducible form allowed for greater convenience in transactions and portability, greater control by the government, but also a greater threat of forgery. The changeover from gold to paper money was an abstraction which separated the monetary value from the material value of the currency itself, a shift also noted by Marx in *Capital* (he even indirectly addressed the problem of the coin as a quantized unit, noting that wear and tear gradually reduces the weight of the coin). This shift allowed for a greater variation in value, with the strange result that the monetary value of the currency could actually become *less* than the material value of the money. In Hungary of June 1946, during some of the worst inflation in history, the 1931 gold pengö was valued at 130 million trillion paper pengös, and in 1923 Germany, the German mark was quoted at four trillion to the dollar; for smaller denominations, neither currency could have been worth the paper it was printed on. The United States Government found itself with a similar problem around 1981, when it decided to use a zinc-copper alloy in the production of pennies instead using all copper, because the price of copper had risen to the point where the penny was worth more as copper than as currency.

The monetary system turned use value into exchange value; with it you *can* compare apples and oranges, or at least their exchange value. The kind of thinking that developed along with quantized value allowed for people to measure and compare someone's "net worth" and use aphorisms like "Time is money". The development of valuation expressed in numerical form became a shorthand for communicating worth, and while it was convenient for purposes of trade, it also promoted a more narrow definition of what "value" was, limiting it to societal norms, and turning whatever it touched into a commodity.

Following the shift to paper money, further abstraction occurred with the use of checks, money orders, and credit cards; these were issued by institutions outside of government, and safer and often more convenient to use than the actual currency backing it up (or supposedly backing it up). This paved the way for the next abstraction, which money is still undergoing; the shift from paper to electronic funds transfer systems. Although it has yet to become the dominant form of common everyday usage, government and business have been using it since 1965.

In his 1968 book *Money in the Computer Age*, F. P. Thomson points out that physical money is easily damaged, destroyed, lost or stolen, subject to counterfeiting, and expensive for governments to manufacture, distribute, keep track of, and eventually replace. The paperwork, administrative staff, security precautions, and other overhead involved in money transfers, as well as postage or other forms of delivery, add to the cost of every monetary transaction. According to Thomson, these transactions are also more cumbersome, more time-consuming and less efficient than electronic versions of funds transfer.[22] Of course, the conversion over to electronic systems is also costly, and is one of the reasons why the changeover is so gradual, and why government and business lead the way.

The replacement of physical forms of money is only one function of electronic money; a more important function is the creation of new forms of money or value on electronic systems, like *electronic daylight money*. These "intraday overdrafts" are electronic interest-free loans given out by the Federal Reserve, which are loaned out in the morning and must be paid back before the end of the day. The speeds at which electronic transactions occur allow huge numbers of transactions to occur in the few hours these loans are available, and hundreds of billions of dollars are lent out in this manner every business day.[23]

Other types of electronic money include bank-linked money flows and corporate barter flows. The bank-linked money flows are not deposit money, but their settlement is in deposit money or cash (credit, cash, or ATM transactions). Although performed electronically, there is still a link to bank deposits. Corporate barter forms, however, are sent electronically between corporations, or internally within a corporation and its various offices around the world. Such transactions are private, do not involve the Federal Reserve or the government, and are difficult even to track. Since they are often not even bank-linked, they abstract money further still.[24] This form of money is so abstracted that it is difficult to separate out from other data. According to Elinor Harris Solomon;

> The electronic money flows can be a part of broader kinds of confidential and proprietary corporate information all "bundled together." There is no way to unbundle the money-like information from that sent along within corporate message flows on a transponder portion of the satellite; and neither the Fed nor public would have any right to obtain the private information if such unraveling of data skeins were possible.
>
> Such money is continuous, not intermittent, and less predictable than daylight money in creation, timing and future scope. Unlike the daylight group we have little information on amount or flow. The size of flows is private information, published infrequently (say, once a year) and without any mandatory requirements or surveillance. The money exists as quite intangible information flows. Much of it moves on private satellite or nonbank links outside the banking system altogether.
>
> Yet messages transmitting such information-based "value" are used just like any other money in payment for goods and services. They can contribute to the volatility of financial asset prices. Such money "buys" financial assets at wholesale, futures, and options of financial assets and commodities; it hedges private traders and investors against risk, flows around the world on private wires currently built and in the process of construction and expansion.[25]

Ironically, in the digital age, we have come full circle and returned to a moneyless barter system. But this barter system has been abstracted far from nature; transactions occur at nearly the speed of light, trades need not refer to physical objects, strongboxes and vaults become data sets, and money becomes information.

Information

Unlike time or space, information is not a continuum that can easily be divided into units. The definition of "information" has changed over time, and it is only in the 20th century that it has become thought of as a precisely measurable entity that can be separated into discrete and indivisible units. In order for this to occur, there had to first be a way to express information numerically, so information (as we know it) made its first steps towards quantization with the development of quantitative thinking.

As counting became important, number systems developed, including the decimal system which has become the basis for almost all modern quantitative thinking. The base ten system, however, is only one possibility; the Yuki of California use cycles of eight, a sexagesimal system once existed in Babylonia and ancient Greece, a quadragesimal system in Latvia, and a vigesimal system in France.[26] And, of course, computers uses base two, or binary, for their operations.

Besides their role as data, numbers were also used to keep track of nonnumeric information. As printed information moved from the clay tablet to the papyrus roll, to the parchment codex and finally to the printed book, page numbers became increasingly important for the locating of printed matter (paragraph numbering had been tried, but was not successful). Certain documents, like the books of the Bible, go even farther and have numbered lines and verses because page numbering would not be consistent between translations. Page numbering, as the most convenient system, continues to be important for academic citing; and in the publishing business, page numbers have even become copyrightable and a source of profit and controversy.[27]

As pages were numbered in books, so collections of books were numbered in libraries. The library began as a record of land ownership, the decrees of kings, laws, genealogies, and religious writings, but in the Middle Ages the book and library came to be seen as valuable, along with the printing press that made many of them possible. For a library to be of use, there had to be a way of locating a desired text; thus indexing and cataloging alphabetically by subject and author began, though not without resistance.[28] Printed library catalogs which gave shelf numbers of books gave way to card catalogs when Melvil Dewey submitted his Dewey Decimal Classification system at Amherst in 1873; other systems like Charles Cutter's rival 35 base Expansive Classification system (using letters and numbers) existed before Dewey's, but Dewey's system won out in the United States. In 1903, the Library of Congress began its classification system, and many libraries subscribed. And now, in addition to call numbers, many libraries use bar coding numbers to keep track of their inventories.

As the number of libraries grew, so did the desire for an international standard of book-numbering, and in 1970 the International Standard Book Numbering (ISBN) system was started. This ten-digit number includes a group identifier indicating the national, language, geographic or other area where the book was published, a publisher prefix indicating a specific publisher, a title number, and a check digit. And since the late 1980s, libraries have used on-line catalog systems which may one day be used to retrieve the texts of whole books. The computer, using an ASCII or Unicode number to represent every letter and punctuation symbol, seems the logical tool for library numbering.

Information and its quantization, however, was not limited to the library. Renaissance thinkers including Galileo, Descartes, Kepler, Huygens, Newton, and others contributed to a mathematicization of science, beginning with physics and astronomy, which became an attempt to mathematize all natural phenomenon. This dream continues today, in such projects as Benoit Mandlebrot's fractal geometry and the field of biomathematics. In the early 19th century, atomic numbers were devised and the periodic table of elements was created, and in the early 20th century, Bohr orbitals described the limited number of possible orbits available to electrons. Max Planck developed the notion of quanta, the indivisible and smallest possible units of energy (along with the supposedly smallest possible units of space and time, mentioned in the preceding sections). To many, it seemed that nature was already quantized, and science was finally able to study and measure its units; all matter and energy were divided into fundamental particles and forces. But the number of discovered and expected particles kept growing, and Heisenberg's Uncertainty Principle revealed the limits of measurability. Einstein's theories of relativity brought an end to notions of absolute space and time. More recent theories, like those of superstrings and Penrose twistors, have moved into dizzying mathematical abstraction. Although newer theories of contemporary physics are less bent on quantizing nature, they still are heavily mathematical and represent things purely numerically, and seem far removed from descriptions of everyday experience.

The success of the hard sciences in mathematizing their fields of study prompted the social sciences to do likewise. In the early 19th century, the English Utilitarians, under the leadership of philosopher Jeremy Bentham, began a series of commissions that collected social statistics regarding public life. These were complied into the Victorian Blue Books, which Karl Marx used to write his indictment of capitalism. Gradually politics came to rely more and more on statistic reports, and were largely reshaped by them.[29] And once politicians found uses for information collection, it was inevitable that the information age would get government support just as science had.

The success of science has largely been a force behind the quantization of everyday life in this century, and scientific method has been used to attempt to quantize and mathematize everything into numbers in a scientific study. The 1890 census was the first census to be tabulated by machine; Hollerith punch cards (discussed in the following chapter) made working with vast amounts of data and statistics possible, and created an information industry in the early 20th century which led to the development of IBM.

In some ways, the new statistical science had the aura of novelty. From 1891 to 1894, over nine thousand visitors to the South Kensington Science Museum paid a fee to have their measurements recorded in Sir Francis Galton's Anthropometric Laboratory.[30] Galton was such a lover of statistics that he reportedly carried a device for recording his estimate of the beauty of women he passed in the street. Since the boom in statistical science around the turn of the century, statistical methods have been applied to many areas of science and industry; Taylorism, for example, attempted to find the most efficient movements by which workers could perform their tasks. Economics and many social sciences have become highly statistical to the point that there are now more jobs for mathematicians than ever —in industry, business, and on Wall Street.

Like time, space, and value before it, information was increasingly abstracted; it became numeric largely because much of it was made up of time, space, and value measurements. Throughout history, there have been changes in what is considered to be information; at present anything numerically expressible counts, since computers cannot deal with anything else directly. A "difference" is often given as a measure and expressed as a number; qualitative differences are reinterpreted and abstracted into measurable quantitative ones.

The ultimate abstraction of information came in 1949, when Claude E. Shannon developed the notion of the "bit" (short for *bi*nary dig*it*), which has two states, on and off, usually represented by one and zero. As the main unit of currency in information science, the bit allowed amounts of information to be measured regardless of content; it looked at all forms of information as a series of choices between different options. The shift from facts to bits as a unit of information leads to greater decontextualization. Strings of ones and zeroes are as abstract and context-dependent as one can get; whereas facts at least have a semantic value, bits by themselves are meaningless; like letters of the alphabet, they are doubly articulate, gaining meaning through context alone. They are the endpoint of abstract representation for everything that has been quantized, and anything that can be numerically expressed.

Growing Number? Quantization and Thought

The quantization of everyday life is so thorough that it is often invisible and even seems inevitable. It can be found in such things as recipes, shoe sizes, and even restaurant visits (*Denny's*' receipts list not only the foods ordered and their prices, but the server, the date, the 'business date' and time of the visit (down to the minute), the table number, and the check number; the visit has its own unique numerical representation). People have even become associated with series of numbers which are used to keep track of them, or to locate them; address numbers, phone numbers, employee numbers, account numbers, ratings of various sorts, social security and other personal identification numbers, and so on. People's minds are evaluated by IQ numbers or test scores. As early as 1934, the FBI was indexing fingerprints, numerically encoding them on punch cards. When the day comes when the human genome and people's DNA are completely transcribable, databanks will be waiting for them, looking for correlations with other numerical representations of people.

The usefulness of quantization in thought is obvious; in many ways it is the basis for Western culture, and certainly for science. On the personal level, the abstracting of things into a series of manipulable units allows for methods of problem-solving and procedural knowledge, which have produced great achievements and insights. But its success can also blind one to its shortcomings and dangers; it is, after all only one cognitive style or way of thinking, and there are others involving more intuition, imagination, or creativity. There are other ways of thinking that see relationships, look at things without seeing divisions and units, and are less goal-oriented. After all, information is hardly the same thing as knowledge, wisdom, ideas, or reason. In her book *Ethnomathematics: A Multicultural View of Mathematical Ideas*, Marcia Ascher examines the place of mathematics in non-Western cultures:

> In Western culture there is a belief that numbers carry a great deal of information. The belief has become particularly pervasive over the last hundred years... Most other cultures have less belief in the value of the information conveyed by numbers. This difference in concern may be associated with technology or population size or the domination of other interests such as spirituality, aesthetics, or human relations. A few native Australian groups, for example, are noted for their lack of interest in numbers. They are also noted for the richness of their spiritual life and overriding focus on human relations. Perhaps most important, however, is that many other cultures value contextual understanding rather than decontextualization and objectivity.[31]

As mentioned above, decontextualization is one of the dangers of digital abstraction. Quantization is inherently reductionist; by its very nature it seeks to find or create divisions, its representations are discrete rather than the continuous. It sees a whole as a sum of its parts, and its abstractions are always simplifications. On this topic, physicist Werner Heisenberg observes:

> We may begin with this question: What is abstraction, and what part does it play in conceptual thought? And we may formulate an answer somewhat as follows: abstraction represents the possibility of considering an object or group of objects under *one* viewpoint while disregarding all other properties of the object. The essence of abstraction consists of singling out one feature, which, in contrast to all other properties, is considered to be particularly important in this connection. As can easily be seen, all concept formation depends on this process of abstraction, since concept formation presupposes the ability to recognize similarities. But since total sameness never occurs in practice among phenomena, the similarity arises only through the process of abstraction, through the singling out of one feature while neglecting all the others.[32]

Abstraction can obscure more than it highlights. When this kind of reductionist attitude is applied to people instead of objects, grave depersonification can occur. Theodore Roszak vividly paints a grim picture of Western culture's numberlust:

> However ambitiously "information" may be defined by its enthusiasts and specialists, all that the data banks and their attendants are after is data at the most primitive level: simple, atomized facts. For the snoops, the sneaks, the meddlers, data glut is a feast. It gives them exactly what their services require. They exist to reduce people to statistical skeletons for rapid assessment: name, social security number, bank balance, debts, credit rating, salary, welfare payments, taxes, number of arrests, outstanding warrants. No ambiguities, no subtleties, no complexities. The information that data banks hold is life stripped down to the bare necessities required for a quick commercial or legal decision. *Do or don't give the loan. Do or don't rent the property. Do or don't hire. Do or don't arrest.* This is human existence neatly adapted to the level of binary numbers: off/on, yes/no.[33]

Digital-style thinking can easily lead to all-or-nothing positions, extremism, stereotypes, and oversimplification. The very notion that there is always necessarily one "best" way of doing things can also be a result of a quantizing world-view. Once things or methods are subject to measurability, there is the possibility to compare them numerically, even though it is an often misleading way of assessing their value. Jacques Ellul, a 20th century philosopher who has examined the relationship between human beings and technology, has written about what he calls "technique", the style of thinking that attempts to find the "one best solution" to whatever problem it encounters, and has maximum efficiency as its greatest goal. As he points out, this can conflict with a democratic system, which offers several options to choose from, tending towards a more totalitarian one in which one "best way" is offered, and which ostensibly holds efficiency as the greatest good. Even the notion of scientific "progress" itself exhibits this teleology; after all, exactly what is being measured when a measure "progress" is taken?

Measurability and exchange value have become too important in the estimation of worth. Perhaps Western society as a whole is guilty of *quantophrenia*, a condition defined as "an obsession with and exaggerated reliance upon mathematical methods or results, especially in research connected with the social sciences."[34] To rely on series of numbers as representations of the world around us is to abstract us from it, distancing or even cutting us off from experience, the subtler sides of things, moods, and feelings that cannot be captured in words, much less in numbers. It is no wonder that Frederic Jameson finds the postmodern era to be characterized by a waning of affect. As numbers grow, we grow numb-er.

Non-western native cultures like the Navajo and the Inuit, which have little or no use for numbers and whose notions of time, space, and value do not involve division into parts or units, show that alternatives to a quantizing world-view are possible and perhaps even desirable. Because quantization involves an abstracting of the real, it also tends to hide mediation, which occurs more easily and readily when representations are simplified (and especially when their basis is numeric). A cultural world-view in which quantization plays a major role was necessary for the growth of digital technology from its precursors to its present form. The quantization of everyday life not only made the development of digital technology *possible*, it made it seem *natural*.

NOTES

1. Hellemans, Alexander, and Bryan Bunch, *The Timetables of Science: A Chronology of the Most Important People and Events in the History of Science*, New York: Simon & Schuster, ©1988, page 3.
2. Turner, Anthony John, *Of Time and Measurement: Studies in the History of Horology and Fine Technology*, Brookfield, Vermont: Variorum, ©1993, page 19.
3. According to Hellemans and Bunch, *The Timetables of Science*, page 10,

> Since the year is not exactly 365 days, the Egyptian calendar went into and out of alignment with the seasons, with a period of about 1455 years. Knowing this, astronomers have speculated that the year of 365 days was instituted around 4228 BC or 2773 BC.

4. Turner, Anthony John, *Of Time and Measurement*, pages 19-21.
5. According to Carlo M. Cipolla,

> In 1841 a petition presented to the Town Council of Lyon stressed the fact that in the town 'is sorely felt the need for a great clock whose strokes could be heard by all citizens in all parts of the town. If such a clock were to be made, more merchants would come to the fairs, the citizens would be very consoled, cheerful and happy and would live a more orderly life, and the town would gain in decoration. ... Thus a combination of civic pride, utilitarianism and mechanical interest fostered the diffusion of the clock despite its relatively high cost.

Cipolla, Carlo M., *Clocks and Culture 1300-1700*, New York: Walker and Company, ©1967, pages 42-43.
6. Guye, Samuel, and Henri Michel, *Time & Space: Measuring Instruments from the 15th to the 19th Century*, New York: Praeger Publishers, ©1970, page 85.
7. Cipolla, Carlo M., *Clocks and Culture 1300-1700*, pages 43-44.
8. Itano, Wayne M., and Norman F. Ramsey, "Accurate Measurement of Time", *Scientific American*, Volume 269, Number 1, July 1993, page 57-59.
9. As an article in *Science* explains;

> The adjustment was necessary because man-made clocks-- which run on frenetic oscillations of the atom-- are superior to the ancient timekeeper, the earth, which is running down. The earth wobbles, lurches, and continuously slows down on its axial spin, demanding a leap second's rest every so often to catch up with atomic time. ... The earth, sun, moon and stars are now held accountable for their movements according to the atom. The earth, in particular, has shown itself to be sadly deficient, losing about one millisecond (a thousandth of a second) per day.

The same regularities of nature by which humanity set its clocks are now themselves under the scrutiny of those clocks; scientists have determined the "correct time" and it is now the earth which is "sadly deficient". From "A Matter of Time", *Science*, volume 238, number 4834, December 18, 1987, page 1461. A leap second was added on December 31, 1987, at midnight; and another was added on June 30, 1992; 23h 59m 59s was followed by 23h 59m 60s and then by 0h 0m 0s, July 1, 1992 (see *Astronomy*, volume 20, June 1992, pages 24-26).

10. Zupko, Ronald Edward, *Revolution in Measurement: Western European Weights and Measures Since the Age of Science*, Philadelphia, Pennsylvania: American Philosophical Society, ©1990, pages 22-23.
11. From *Science and Technology Desk Reference: 1,500 Answers to Frequently-Asked or Difficult-to-Answer Questions*, edited by the Carnegie Library of Pittsburgh Science and Technology Department, under the direction of James E. Bobick with the assistance of Margery Peffer, Washington, D.C.: Gale Research, Inc., ©1993, page 630.
12. From *The 1994 Information Please Almanac*, 47th ed., Otto Johnson, Executive Editor, Boston and New York: Houghton Mifflin Company, ©1994, page 426.
13. Oxford Reference, *A Concise Dictionary of Physics*, Oxford and New York: Oxford University Press, ©1990, page 212.
14. Newton, Norman T., *Design on the Land; The Development of Landscape Architecture*, Cambridge, Massachusetts: Belknap Press of Harvard University Press, ©1971, page 6.
15. Since the largest degree of meridian is 69.172 miles, at the equator, the widest possible arc second of meridian would be 101.4522667 ft., and the smallest, at the poles, would be zero. The idea of precise global coordinates have created an industry of global positioning satellites and the devices which can pinpoint the user's position, often down to a few meters.
16. Stilgoe, John R., *Common Landscape of America, 1580 to 1845*, New Haven, Connecticut: Yale University Press, ©1982, page 99.
17. Ibid., page 105.
18. Ibid., pages 106-107.
19. "The first two digits of the "+4" code denote a delivery "sector", which may be several blocks, a group of streets, a group of post office boxes, several office buildings, a single high-rise office building with multiple firms, a large apartment building or a small geographic area. The last two digits denote a delivery "segment", which might be one floor of an office building, one side of a street between intersecting streets, a firm, a suite, a post office box or a group of boxes, or another specific geographic location." From *California Zip +4, State Directory*, U. S. Postal Service, page P-8.
20. From Ray B. Westerfield's "Money" entry in *Collier's Encyclopedia, Volume 16 Metal to Musial*, Lauren S. Bahr, Editorial Director, and Bernard Johnston, Editor in Chief, New York: P. F. Collier, Inc., ©1993, page 442a.
21. According to *The Guinness Book of World Records, 1994*, New York: Sterling Publishing Co., ©1994, page 140, the earliest coins were electrum staters of King Gyges of Lydia, Turkey, c. 670 BC; a footnote to this adds that Chinese uninscribed "spade" money of the Zhou dynasty has been dated to 770 BC. Paper money was first tried in China around 812 AD, and became prevalent around 917 AD.
22. Thomson, F. Paul, *Money in the Computer Age*, New York and Oxford: Pergamon Press, ©1968, pages 55-105.

23. Elinor Harris Solomon, "Today's Money: Image and Reality", in *Electronic Money Flows: The Molding of a New Financial Order*, edited by Elinor Harris Solomon, Boston, Massachusetts: Kluwer Academic Publishers, ©1991, pages 21, and 34-35.
24. Ibid. Solomon writes on page 27,

> Corporate barter flows may serve the needs of different firms as buyers and sellers interact in payment with one another (the "interfirm flows"). Or they may cross oceans within the corporate infrastructure so as to place buy-and-sell orders with speed, or to distribute products and parts electronically within the infrastructure (the "intrafirm flows"). . .
>
> As you get farther and farther away from the reserve base and from the money that passes through banks, it becomes harder to track down the "money." This is especially true if there is bundling of a lot of different kinds of corporate information or settlement in some kind of commodity account (oil or futures, for example). The value of an aggregate flow gross becomes supported by a dwindling reserve and money base relative to expanded electronic blob as a whole. Like the creatures beyond the time warp of Arthur Clarke's *A Space Odyssey: 2001*, [sic] the physical links may be severed altogether for the Group D money forms.

25. Ibid., page 25.
26. On the Yuki, see Ascher, Marcia, *Ethnomathematics: A Multicultural View of Mathematical Ideas*, Pacific Grove, California: Brooks/Cole Publishing Company, ©1991, page 9. On the other systems, see Kula, Witold, *Measures and Men*, page 306, notes 2, 3, and 4.
27. Page numbering has allowed West Publishing an unfair advantage over competitors;

> Of course, the actual words of legal decisions are public domain. They cannot be copyrighted. What West Publishing of Eagan, Minnesota has managed to acquire is ownership of the quasi-official page numbers of federal decisions. Many judges recommend or demand that lawyers appearing before the court include such citations to specific West-owned publications, such as the *Supreme Court Reporter*, the *Federal Reporter*, *Federal Supplement*, or *Federal Rules Decisions*. Under a copyright claim that has survived at least one challenge in court, other publishers of legal decisions are not permitted to show, in the margins of the books or in headers of their databases, parallel citations that describe where decisions appear in West-owned books or in the West-owned database, Westlaw. This puts West in the catbird seat: Any publisher is free to compile collections of federal decisions, but without citable page numbers these collections are little more than worthless.

From "Who Owns the Law?" by Gary Wolf in *Wired*, May 1994, page 98.
28. See Johnson, Elmer D., *History of Libraries in the Western World, Second Edition*, Metuchen, New Jersey: The Scarecrow Press Inc., ©1970, pages 487-494; and Metcalf, John Wallace, *Information Retrieval, British and American, 1876-1976*, Metuchen, New Jersey: The Scarecrow Press Inc., ©1976, page 19.
29. Roszak, Theodore, *The Cult of Information: A Neo-Luddite Treatise on High-Tech, Artificial Intelligence, and the True Art of Thinking*, Berkeley and Los Angeles, California: University of California Press, ©1994, page 158.
30. Eames, Office of Charles and Ray, *A Computer Perspective: Background to the Computer Age*, Cambridge, Massachusetts and London, England: Harvard University Press, ©1990, pages 28-30.
31. Ascher, Marcia, *Ethnomathematics: A Multicultural View of Mathematical Ideas*, page 6.
32. Heisenberg, Werner, *Across the Frontiers*, New York: Harper and Row, Publishers, ©1974, page 71.
33. Roszak, Theodore, *The Cult of Information: A Neo-Luddite Treatise on High-Tech, Artificial Intelligence, and the True Art of Thinking*, page 211.
34. From *The Oxford English Dictionary, Second Edition, Volume XII: Poise-Quelt*, page 980.

2.
Digital Technology Develops

The quantization of everyday life set the stage for the development of digital technology. The turn-of-the-century expansion of statistical science meant vast opportunities for information-processing industries and investment in their technologies. Other forces such as two World Wars, military spending, and advances in science and engineering also contributed, providing goals and funding which would direct the technological development that resulted by mid-century in the birth of the computer age. Through innovation and diffusion, much technology initially designed for military use spread into the areas of art and communication. The ideas behind the technology, however, come from a variety of sources stretching far back into history through which we can trace the digital computer to its beginnings.

While this chapter attempts to trace the development of digital technology, it is *not* in any way intended as a history of computers nor of the computer industry, though it touches on both. Nor is it a technological history; rather, it is a look at how the concept of "digital" was developed into an important technology, its primary uses, and some of the cultural implications. It examines several areas; first, the mathematical background, binary numbering in particular, which provided the theoretical basis for modern digital technology. Second, the evolution of inventions in which a series of discrete, simple presences and absences (for example, beads, nails or holes) are recorded and stored as data to be read again later. Finally, the last section looks at the combining of theoretical and practical threads during the development of the digital computer. Before moving to these sections, it is useful to pause to consider two philosophical notions upon which the concept of digital is based.

Since the days of Aristotle, the twin notions of *presence* and *absence* have haunted Western philosophical thought. Eastern philosophy had the yin and the yang, but these were said to be present within each other to some degree, unlike the Aristotelian notions of *A* and *Not-A*, truth and falsehood, or other dichotomies (on/off, yes/no) whose two sides were mutually exclusive and non-overlapping. Of course, presence and absence rely on each other for their definition, as each implies the possibility of the existence of the other. The theme of presence/absence has been at the heart of many philosophical debates, underlying arguments of existence and nonexistence, substance and form, and epistemological criteria in general. A wide range of contemporary thought, including fuzzy logic and the writings of Jacques Derrida, has sought to break down the rigid Aristotelian division between presence and absence. But still, in everyday life at least, the usefulness of this division has continued and its ingrained influence has led into an even greater dependence on it. Today, the concept of presence/absence is closely associated with binary numbering, most often iconically referred to as one (presence), and zero (absence).

The concept of "1" existed from the time the first tally marks were made, as its form, a short line, can attest. A "1" is abstract by itself, and when it is applied to objects or things it serves to separate them out from their surroundings, setting them off as discrete entities that can be counted.[1] As one finds when comparing the perfect circle of mathematics to the imperfect ones actually found in the world, mathematics is made of idealizations that the material world can never fully realize. Whenever a one-to-one correspondence between them is attempted, abstraction occurs and quantization begins (the term "one-to-one" is perhaps even inappropriate here, as both ends of it are already abstract).

The concept of *zero* is, in some ways, more abstract than that of *one*; it is a presence (of a signifier) which signifies an absence of "something" which is not present. The zero came about because a place-holder was needed in base positional number systems to signify an empty place (in base ten, for example, "102" means one hundred, no tens, and two ones; without the zero, the number would be mistaken for "12"). The zero first appeared in the third century BC in the Babylonian number system, and is found in sixth century AD Sanskrit numeration, but it was not until the twelfth century that the zero sign was introduced into the West. The introduction of the concept of zero into the West was nothing less than "a major signifying event" according to Brian Rotman, whose book, *Signifying Nothing: The Semiotics of Zero*, begins:

> The first purpose of this book is to portray the introduction of the mathematical sign zero into Western consciousness in the thirteenth century as a major signifying event, both in its own right within the writing of numbers and as the emblem of parallel movements in other sign systems.
>
> Specifically, I shall argue that certain crucial changes in the codes of number, visual depiction, and monetary exchange that occurred as part of the discontinuity in Western culture known as the Renaissance - the introduction of *zero* in the practice of arithmetic, the vanishing point in perspective art, and imaginary money in economic exchange - are

> three isomorphic manifestations, different, but in some formal semiotic sense equivalent models, of the same signifying configuration... In all three codes the sign introduced is a sign about signs, a meta-sign, whose meaning is to indicate, via a syntax which arrives with it, the absence of certain other signs.[2]

One and zero, then, form the roots of mathematical abstraction from which the Renaissance, the Industrial Revolution, and the Computer Age gain their impetus. Binary numeration, consisting solely of zeroes and ones, was discovered around the time of the Renaissance, further developed during the Industrial Revolution, and came into full maturity and use during the Computer Age.

Binary Numbering

Zero and one form the basis of binary numbering, the base two system composed of two numerals (zero and one) instead of the decimal system's ten. Although the ancients had systems that relied on dividing things in half, it was later, around the time of the Renaissance, that number systems other than base ten were investigated in the West. Thomas Hariot (1560-1621) conceived of numbering systems based on adding powers of two, but never published his work; Francis Bacon developed a "bi-lateral" code for the letters of the alphabet, using A and B instead of 1 and 0; and the I-Ching of the Orient had a binary basis. The first *publication* of work on binary arithmetic was by Bishop Juan Caramuel Y Lobkowitz, in 1670. His work was ignored, unfortunately, and so discovery of the binary system is often attributed to Gottfried Wilhelm Leibnitz (sometimes spelled "Leibniz"), who published his "Explication" 33 years later in 1703.[3]

As an alternate numbering system, binary was little more than a mathematical curiosity to nonscientists and of interest only to mathematicians; it would only become useful to science after further investigation. Leibnitz's work with binary was more in-depth than that of his predecessors; he described addition, subtraction, multiplication, and division in binary. In 1697, in a letter to the Duke of Brunswick, he explained the notation system, which he believed reflected creation and the origin of numbers, since God (unity or one) created all things out of nothing (zero). In his letter he gave examples of the ease of binary addition and multiplication, but added, "However, the intention is not to use this manner of reckoning in other than the study of and search for the secrets of numbers, and not at all for use in every-day life."[4]

In the next few centuries after Leibnitz, binary numbers remained only a subject for mathematical study. The next advancement that found a use for binary came around the middle of the nineteenth century, in mathematician George Boole's invention of logical operators, a "mathematics of logic" whose rules govern logical functions rather than numbers. These operators, AND, OR and NOT did not necessarily use binary numbering, but they were binary by nature, dealing with two possible states, *true* and *false*.[5] After Boole published his work, a number of logic machines were built using his principles (such as Charles Earl Stanhope's Demonstrator, Jevon's Logical Piano, and Allen Marquand's logic machine). These machines mechanized deductions based on syllogistic reasoning, and some were said to calculate them faster than a person could. Previously, formal logic had been considered a philosophical discipline, and it was Boole's work that helped to formalize it into a branch of mathematics, reviving and creating new interest in it. Binary was now underlying the basis of logic machines, but it was not until the twentieth century that the connection would be fully exploited.

In 1937, Claude E. Shannon's Master's thesis at MIT, "A Symbolic Analysis of Relay and Switching Circuits" connected the ideas behind Boolean operators with the design of electronic circuitry and switching, providing the theoretical basis for digital computers. Although his work depended on that of others and he was not the first to make the connection, his treatment of it was the first to receive wide attention.[6] Once the connection was made, it was clear that binary was the simplest system to use, the two states represented by on and off. In 1948, Shannon wrote: "The choice of a logarithmic base corresponds to the choice of a unit for measuring information. If the base 2 is used the resulting units may be called binary digits, or more briefly *bits*, a word suggested by J. W. Tukey."[7]

The bit, as a simple choice between two alternatives, is the smallest possible unit of information, and it became the basis for modern digital technology; it is the essence of the presence/absence dichotomy mathematized. All information can be stated as a choice made between a number of alternatives; even a letter of the alphabet can be seen as one choice out of 26 possibilities. These alternatives can be elaborated into a branching series of choices, each of which has two alternatives; and each of these is then represented by bits set at either one or zero. Thus, for n number of bits, there are 2^n possibilities.

The impact of the concept of the bit was great and continues to be, within information science and in other areas. For example in physics, quantum physicist John Wheeler suggests that "The basis of reality may not be the quantum, which despite its elusiveness is still a physical phenomena, but the bit, the answer to a yes or no question, which is the fundamental currency of computing and communications."[8] Whether or not the bit is really "the basis of reality", it certainly has been a useful concept both in theory and its practical applications over the years. Even before it was formally recognized, the concept behind the bit, in which information is stored as a series of presences and absences, had been developed along several lines, and it is to these inventions that we now turn.

Early Digital Inventions

The mathematical basis for digital technology was slow to develop compared to the evolution of inventions able to read (and later record) data represented as presences and absences in the form of beads, nails, or holes arranged sequentially in "readable" media like discs, cylinders, punched cards, or paper rolls. Early digital inventions developed concurrently in three main areas; mechanical calculating devices, means of controlling an instrument or machine, and means of communication. The computer is a logical antecedent of these functions, bringing all three together and combining them with the mathematical background described above.

The earliest calculating devices were little more than counters ("calculus" is the Latin word for "pebble"), or the fingers of the person doing the counting. As larger numbers were needed, devices like the abacus came into use, along with other operations such as multiplication and division. The first mechanical calculator, Wilhelm Schickard's Calculating Clock, was built in 1623. According to his surviving sketches, it bore a slight resemblance to an early computer console, displaying its numbers behind notches in strips of wood aligned in vertical rows above a bank of dials set across the bottom. Series of gears made borrowing and carrying automatic, and a bell notified the user when a number was beyond the machine's six-digit capacity.[9] Because of the Thirty Years War, the Calculating Clock was forgotten, while other mechanical calculators came about; Blaise Pascal's Pascaline in 1642, Samuel Morland's adding machines around 1672, René Grillet's of the same period, the Liebniz calculator completed in 1674, and a host of others. All these early calculators performed their functions by a series of gears, cranked by the operator, who also had to manually record the results. It was not until Muller's difference engine, in 1786, that a device was planned that could print out its results on paper. The idea behind the difference engine was the "method of differences" used in its calculations, a shortcut developed by makers of mathematical tables. Instead of calculating each member of a series individually, it used the differences (or second or third set of differences) between terms. For example, the differences between squares (1, 4, 9, 16, 25, 36, etc.) turn out to be consecutive odd numbers (3, 5, 7, 9, 11, etc.), and the differences between each of those terms is always 2.

Of all the people building or attempting to build difference engines, the man most associated with the difference engine was Charles Babbage, whose designs were the most ambitious. Michael R. Williams describes one design:

> The entire mechanism would have been about 10 ft high, 10 ft wide, and 5 ft deep. It was to be composed of seven vertical steel axles, each of which would carry 18 brass wheels about 5 in. in diameter. Each vertical axle would represent one of the six orders or differences while the seventh axle would store the value of the function being computed. These values were represented on the axles by the positions of the 18 brass wheels. Each wheel was engraved with the digits 0 to 9 around its circumference and thus, by simply turning the wheels so that they displayed their various digits, any 18-digit number could be represented on each vertical axle.[10]

Babbage's difference engine was never completed, though years of labor and thousands of pounds sterling had gone into it. During the long period of stops and starts while working on it, Babbage had conceived of an even more powerful machine, the Analytical Engine, which would have been the first general-purpose programmable computing machine, running programs stored on a series of punched cards and performing complex functions using variables. It was designed to have around 200 individual columns of gear trains and number wheels, it would be capable of printing, plotting, or punching its results onto paper, and it could even alter its courses of action, using its results as feedback. Babbage even thought of powering the engine with electricity, but as electricity was not yet well understood, he abandoned the idea.[11] Due to the difficulties in making the difference engine, the Analytical Engine was planned mainly as an academic project, existing only as a series of detailed plans.

Although Babbage's engines were the first *calculating* devices to use it, digital technology as a means of machine control had been around for centuries. In his book on the history of the player-piano, Arthur W. J. G. Ord-Hume notes that,

> Long before the first pianoforte was made, the art of programming music for automatic playing was fairly well developed. The pinned wooden barrel represented the highest development of the art and its use and capabilities were understood at least as early as the eighth

century AD. And the oldest surviving barrel organ still in playing condition is dated 1502.[12]

The earliest programmable instruments were controlled by a barrel which had rows of nails across it. Each row was read in parallel to produce a chord; and as the barrel turned, these rows were read in series, producing music. Other later music machines used a large metal disc with a series of perforations in it, which functioned in a manner similar to the pinned barrel. The most familiar use of this kind of technology remaining today is found in wind-up music boxes; but they are not very good examples of what the technology was capable of. As Ord-Hume writes,

> It is a sobering thought that almost every one of the so-called orchestral musical instruments was automated by one process or another between the years 1500 and 1930, from the saxophone of the dance band to the harp of the symphony orchestra, from toy trumpet to banjo, from xylophone to violin. Even the drum could play tattoo or roll from a perforated disc within.[13]

While the pinned barrel and perforated disc were simple and efficient ways of "storing" a piece of music, they were not really recordings; to make them, a craftsman would have to pin or perforate according to a preconceived plan, and barrel-pinning was a carefully guarded skill passed on to one's children. More than just the transcription of musical notes was involved; effects such as the operation of dampers were also encoded on the barrel (both the notes and the effects instructions were stored in the same medium in much the same way data and commands are in the stored-program computer). But the premeditated nature of pinning left little room for improvisation; for real-time digital recording to come about, the player piano would have to adapt technological advances made during the development of the loom.

During the Industrial Revolution, the mechanization of repetitive processes was being attempted in a number of areas, including weaving. In *The Book of Looms*, Eric Broudy describes the development of the mechanized loom:

> The first significant contribution is generally credited to Basile Bouchon, a Frenchman, who in 1725 invented a device for selecting automatically which simples to pull. Cords of the simple were threaded through eyes in a row of needles that could slide in a box. Paper, perforated accorded to the desired pattern, was passed around a perforated cylinder that was pushed against a box containing the needles. Those needles that slid through the holes remained still, while the others, which hit unperforated paper, were pushed back, along with

> the cords attached to them. The selected cords were then pulled down by a foot-operated comb that engaged beads attached to the cords. The cylinder of paper was rotated with each pick of the shuttle, and a new set of holes selected the cords for the next pattern shed.
>
> A few years later M. Falcon improved on Bouchon's invention by adding several rows of needles and replacing the perforated paper with perforated cards linked edge to edge. Each card represented the selection of needles for one shot of the weft. This simplified pattern changing, but the cards still had to be pressed against the needles by a hand-held perforated platen.[14]

The move from a perforated paper roll to punched cards allowed the cards to be reused in different combinations and lent flexibility to the designs, since the cards allowed them to be changed more easily. The design was further improved in 1745 by Jacques de Vaucanson, who put the selecting box at the top of the loom, and attempted to automate the cards further, so that no hand-holding of the platen was needed. Some time later, around 1804, the design was improved again by Joseph Marie Jacquard, now commonly credited with having invented a punched-card controlled loom, even though it originated over twenty years before he was born. Thus the "Jacquard Loom" is misleadingly named, since it was actually a device mounted on top of a treadle loom, and not itself a loom. The fact that it could be added to existing looms helped it sell, and so the name Jacquard became associated with the invention because it was after the addition of his money and time-saving improvements that it came into widespread use.

Stored as they were on hundreds of punched cards, images created on the Jacquard machine were probably the first digital images. As an imaging device, Jacquard's machine was quite sophisticated, and the images produced on looms using it were not at all the blocky, low resolution graphics that one might expect; they were capable of being as finely detailed as a photoengravature. London's Victoria and Albert Museum owns a woven silk portrait of Jacquard from 1839 which, according to *The Cyclopaedia of Useful Arts from 1862*, has "1,000 threads in each square inch in both warp and weft; 24,000 cards were required for the pattern, each card being large enough to receive 1,050 holes".[15] The portrait demonstrates the textures, fine detail, and subtle shadow and lighting effects that were possible. If each Jacquard punched card used for the portrait had 1050 possible holes, each card held 1050 bits —slightly over an eighth of a kilobyte of information. An image requiring 24,000 cards, then, would be 3.15 megabytes of information —not bad for a woven image!

The idea of punched-card machine control was not lost on Charles Babbage, who acknowledged the Jacquard loom as the basis for the

Analytical Engine's punched card system. Thus it was in Babbage's engines that two developing threads of early digital technology, calculating devices and machine control, were brought together.

As mentioned earlier, punched card technology was applied to player pianos as well, and came into success in the early 1880s when it replaced the wooden barrel in automatic mechanisms. The switch from pinned barrels to paper rolls was an important turning point for the player piano; since paper could be perforated easily, reproducing pianos capable of recording a performance in real time became possible, and many famous composers of the day, including Debussy, Gershwin, Grieg, Mahler, and Rachmaninov, recorded performances of their work on the reproducing piano, and some of these rolls survive today. These performances, made around the turn of the century, were the first digital recordings. As Ord-Hume points out, the piano-roll recordings were better than any of the recordings made on analog devices of around the same time; unlike Edison's wax cylinders, piano rolls were not limited to four minutes, they picked up no background noise, and because they were used to operate the piano mechanically, they could reproduce the entire range of the piano rather than the very narrow dynamic range of sound that the analog media of the time could reproduce. While the punched holes fixed the performance indelibly on paper, the replaying of the roll, however, depended on the condition and tuning of the piano used, and so performances of the rolls could still vary.

The 1880s had another problem to which punched-card methods would be applied; the upcoming 1890 census. The U. S. population had grown so large that the 1880 census was still being tabulated in 1887, and taking further growth into account, it appeared the 1890 census would take even longer, perhaps even overlapping the 1900 census. The Census Office held a competition in order to find the best means of mechanical tabulation, and there were three entries; William C. Hunt's colored cards, Charles F. Pidgin's color-coded paper "chips", and Herman Hollerith's electric tabulating machine which used punched cards. The first two systems involved hand sorting and counting, while Hollerith's automated the recording and sorting of data.

Hollerith's first tabulating system used a roll of punched tape with holes in it wound over a metal drum. Metal brushes passed over the tape, and while passing over a hole they made electrical contact with the drum, converting the holes into the presence or absence of current. But, like the shift that occurred in loom technology from Bouchon's paper roll to Falcon's series of cards, Hollerith redesigned his system to use cards to facilitate random access and easier sorting. Needless to say, Hollerith won the competition, his system tabulating around ten times faster than the others.

Hollerith's punched card system, combining digital and electrical technology, was the first large information processing system, and an enormous success. Operators could process thousands of cards a day, and entry of the data was automatic and did not involve hand-written numerals. The census bureau completed a tally only six weeks after the count had begun, and everyone was amazed. The publicity given the success of Hollerith's machines helped them spread fast. Other countries ordered them to use for their censuses, and private industry began renting them as well. An efficient means of handling large amounts of data encouraged the development of large databases, and in this sense, Hollerith's machines helped promote the rise of the statistical sciences.

Hollerith's Tabulating Machine Company grew rapidly, through mainly commercial work. His health failing, Hollerith stepped down in 1911 and allowed his company to be acquired by Charles Flint, who combined the company with three other smaller ones to form the Computing-Tabulating-Recording company (CTR). In 1914, Thomas J. Watson became president, and under his leadership, the company tripled in size, and in 1924 changed its name to International Business Machines (IBM). IBM, of course, flourished and continued to expand, using and refining punched card technology into the late 1960s, when electronic systems became reliable and inexpensive enough to take over the market.

Punched cards remained in use for so long because of their durability, interchangeability, and ease of sorting. As the punched card evolved, more and more data was squeezed onto it; Hollerith's Tabulating Machine Company cards had 45 columns of ten possible punches each, for a total of 450 bits; early IBM cards had 80 columns of 12 punches (960 bits); and Remington Rand cards had 90 columns of 12 punches, for a total of 1080 bits, or 135 bytes; a little over an eighth of a kilobyte per card (nearly the same amount as Jacquard cards, oddly enough). Like paper rolls, punched cards were not susceptible to erasure in magnetic fields, and the cards could be read even without a machine, with a key indicating what the punches represented.

Punched paper tape also continued to be used in certain areas, such as machining, into the 1970s and early 80s, outlasting punched cards by over a decade. Most punched cards and tape were eventually replaced by magnetic tape, which was first used in the German Magnetophone of 1935 and came into widespread use after World War II. Magnetic tape could be read faster, could store more data per square inch than punched tape, and the signal it encoded could be erased and rewritten, making it more flexible than a nonerasable medium. Magnetic discs later allowed storage with greater random access capabilities.

With the appearance of the compact disc in the 1980s, digital recording technology has returned to its roots, and it is amazing how certain concepts have remained the same. The CD stores its 'ons' and 'offs' as a series of pits on the surface of a reflective disc, and thus still bears a slight resemblance to the large Victorian metal discs scored with rows of small perforations, which turned slowly behind a glass window in an upright wooden cabinet, producing the quaint, shimmering music-box sounds of a bygone era.

Developing concurrently with calculating devices and machine control was the third strand of digital technology, involving communication. The simple present/absent nature of the digital signal meant it was less susceptible to corruption and noise, and could be transmitted by uncomplicated means. Series of stations could also receive and retransmit messages without needing to decode and reencode them, resulting in data networks stretching miles over the countryside. The first such networks are over three thousand years old;

> According to inscriptions found on ancient bone and shell artifacts, messages were transmitted acoustically during the Shang Dynasty (1800-1100 BC) by the beating of drums. Almost 300 years later during the Zhou Dynasties (1100-474 BC), the mode of transmission was improved and the first free-space-propagation optical-communications network was established. This network used directional beacon fires for optical emitters. . . .
>
> The resulting optical communications system (beacon fires by night and smoke signals by day) proved to be vastly superior to the beating of drums both in terms of speed and intelligibility. The use of optical communications systems became widespread and persisted, at least until well after the building of the Great Wall, which occurred during the period 403-221 BC. In fact, the Great Wall was constructed to enhance the optical communications system by the deployment of a series of windowed signalling towers along its entire length. This network was built to facilitate military communications but it also proved useful for more mundane messaging. Written national standards existed for the proper manning and operations of the communications network.[16]

The length and extent of the Great Wall suggest the possibility that the network may have been nationwide. In the West, such digital networks would not develop until the late eighteenth century.

Optical telegraph systems, similar to the Chinese ones in purpose though involving different technology, were built in the 1790s by Claude Chappe, a clergyman in France, and by Abraham Niclas Edelcrantz in Sweden. Chappe's signaling device was made up of a long arm with two shorter arms on either end, all three of which could be rotated to several positions, similar to semaphore. The variety of positions meant that 92 different signals could be sent, and these were coded in a book with 92 entries on each of its 92 pages; thus two signals could select one of the 8,464 entries in the book. Edelcrantz's telegraph used a series of ten shutters set in a horizontal or vertical position, resulting in a binary system capable of 1,024 different signals.

According to Holzmann and Pehrson, by 1799 there were roughly 150 optical telegraph stations in use. They were set up across the length and width of France, out to the Spanish border and the Mediterranean, and several points along the shore of the English Channel. In the north, several lines extended into Germany, Belgium, and the Netherlands, while in the south one branch of stations extended across Italy all the way to Venice. The Swedish signal towers were built on the mainland and islands along the Swedish shoreline. By 1852, the optical telegraph had reached its peak; 556 stations covering 4,800 kilometers of lines, linking 29 of France's largest cities to Paris.[17]

In the United States of late 1837, where Native American tribes had their own systems of smoke signals, a new form of telegraphy was being developed by Samuel Finley Breese Morse, who invented the electromagnetic telegraph in 1837 concurrently with Sir William Fothergill Cooke and Sir Charles Wheatstone who invented one in Britain around the same time. The optical telegraph declined as it was replaced by the electromagnetic telegraphic, which did not require a visual connection between stations; it worked equally well at night, in blizzards or in fog, and over much longer distances. As it did not need a series of towers to connect distant locations, it was more economical, especially when one considers that the French optical telegraph stations each employed six workers, totaling about 3000 for the entire country.

The electromagnetic telegraph spread rapidly in America, since there was no optical telegraph system already in place; countries with an optical system were slower in changing over (in France, the transition was gradual; the electromagnetic signals were first used to drive a small semaphore signaling device which displayed Chappe's signals). Unlike Edelcrantz's mirror system or others with multiple signaling elements, the electrical pulses used in the new telegraph were sent in series, and were identical; current either flowed or it didn't, and Morse had to devise a coding scheme to encode letters into these absences and presences of current. Although Morse code contains three basic elements (the dot, the dash, and the 'silence' or lack of current between pulses), it is still essentially binary; the dash is merely a pulse three times as long as a dot. Silences of different lengths between pulses indicate spacings between letters and words. Using ones to represent pulses and zeroes for the intervening silences, Morse code messages can easily be represented in binary form. In choosing dot and dash correspondences for the letters of the alphabet, Morse had efficiency in mind, giving frequently used letters shorter representations; E was a single dot, while T was a single dash, and A, I, M, N, and L each used two symbols (A= dot dash; I= dot dot, M= dash dash, N= dash dot, and L was a long dash). The rest of the letters used three or more pulses. This method is somewhat similar to modern strategies of data compression, and could even be seen as an early form of Huffman encoding, which operates on a similar principle.[18]

Morse code was also the first instance of the electronic transmission of information. All the signals sent had to be encoded and decoded by hand, limiting the speed and accuracy of message-sending. Afterwhile, teletypes to automatically encode and decode alphabet letters were invented, and only later was there any means of storing telegraphic transmissions or redistributed them. Conceptually, these inventions were the predecessors of the telephone, modem, and FAX machine. Morse code became the *lingua franca* used for communication in media including the Aldis Lamp (a device invented for signaling from surface vessels to airplanes in the first World War), wireless telegraphy, and radiotelegraphy; and Morse code is still learned and used by ham radio operators today.

In 1819, eighteen years before Morse code, another form of digital communication was invented which did not transmit its messages visually or aurally. Night warfare made the transmission of commands difficult, due to noise and darkness, leading one of Louis XVIII's artillery captains, Charles Barbier, to develop a means of communication based on dots and dashes in relief on cardboard, which could transmit a message through touch. Barbier later developed the idea into writing for the blind, calling it "Sonography".[19] In 1820, embossed pages of Barbier's coded dots and dashes made their way into the National Institute of the Blind in Paris, where eleven-year-old Louis Braille was studying. Four years later, Braille had simplified Barbier's rather complicated system, reduced the coding down to cells of six dots (three rows of two dots), each of which could represent a letter or numeral. For the first ten letters of the alphabet, Braille devised patterns using the first two rows of dots, while the bottom row was empty. The next ten letters repeat the patterns of the first ten, only with a dot appearing in the first position of the bottom row, and some of the remaining symbols use both dots of the bottom row. The repetition with additional dots added below aids in the identification of letters, and is similar to binary numbering as well. Braille, himself a musician, also developed a means of encoding musical notation with his system, and twentieth century additions to his system include diacritical marks and mathematical notation. As a means of communications using the presences and absences of dots to be read by the fingers, Braille is a form of "digital" communication in both senses of the word!

One other form of communication worth mentioning is the use of the light bulb as an on/off device, and its use in digital imaging. Grids of electric lights were used in the 1890s for advertisement displays and signage, spelling out words and sentences. These signs were especially effective at the time, due to their novelty, and similar ones are still in use today; some of them scroll words and messages across grids of light bulbs, while others such as stadium scoreboards, use large banks of lights for live or instant replay video images.

Light bulbs and banks of them were also commonly used as output devices for early computers, indicating the 'on' or 'off' status of a number of bits. This use can be traced back to 1937, when Bell Labs technician George Stibitz used two flashlight light bulbs in a binary adder he built on his kitchen table. As he recalled later in *Datamation 13* (1967);

> I had observed the similarity between the circuits' paths through relays and the binary notation for numbers and had an idea I wanted to work out. That weekend I fastened two of the relays to a board, cut strips from a tobacco can and nailed them to the board for input; bought a dry cell and a few flashlight bulbs for output, and wired up a binary adder. I wired the relays to give the binary digits of the sum of two one-digit binary numbers, which were entered into the arithmetic unit by pressing switches made of the metal strips. The two-flashlight-bulb output lighted up to indicate a binary 1 and remained dark for binary 0.
>
> I took my model into the labs to show some of the boys, and we were all more amused than impressed with some visions of a binary computer industry. I have no head for history. I did not know I was picking up where Charles Babbage in England had to quit over a hundred years before. Nor did it occur to me that my work would turn out to be part of the beginning of what we now know as the computer age. So, unfortunately, there were no fireworks, no champagne.[20]

And so the light bulb, used as a means of illumination and communication, entered into the computer age, its on-or-off status used for representing binary numbers.

The digital computer was a result of the combination of the three early uses of digital technology outlined here; calculating devices, machine control, and communication. Each of these areas of technology had economic forces driving its developments; apart from their use by mathematicians, the main use of calculating devices was by merchants for trade; machine control was mainly used by industry; and communications were often developed for military purposes (even Braille had its roots in the "night writing" of the French Army). Given these uses of early digital technology —and that these three areas provided its main financial support— it is not surprising that much of computer technology is put to economic, industrial, or military use. Only once the technology was already sufficiently developed did it find its way into everyday life and a variety of other uses.

Into the Computer Age

Babbage's Analytical Engine, had it been built, would have been the first invention to combine the three strands of digital technology. Babbage even foresaw how computers would change the field of mathematics, requiring it to adapt, writing;

> ... at the time that logarithms were invented, it became necessary to remodel the whole of the formulas of trigonometry in order to adapt it to the new instrument of calculation: so when the Analytical Engine is made, it will be desirable to transform all formulas containing tabular numbers to others better adapted to the use of such a machine.[21]

Babbage was right; every function, down to addition and subtraction, had to be translated into binary terms as computation shifted to digital computers which quickly surpassed their analog precursors.

The meaning of the term "computer" itself has changed over time; before World War II, it referred to the human being doing the calculations. At present, the term refers mainly to digital computers, and to a lesser degree, to their analog ancestors. Analog computers existed before digital ones, and have a rich history of innovation. Analog devices include the Astrolabe, the Antikythera device (a gear-driven mechanism for predicting the motions of the sun and moon from the early Christian era), tide predictors from as early as the 15th century to the more modern ones of the early 20th century, and differential analyzers used for calculating ballistic firing tables (from James Thomson's in late 19th century to Vannevar Bush's in the 1930s). The wheels and gears on these devices did not have a discrete number of positions that they could occupy, as did the gears in the Analytical Engine. Analog machines had some advantages over digital ones both in terms of construction, as well as for problem solving in areas like integration or astronomy. Digital machines could only take over analog functions when a mathematical understanding of the phenomena in question was reached, and when the units they measured were small enough to match the degree of precision possible with analog devices.

The question of which machine can claim to be the "first" digital computer depends on your criteria and definition of what a "computer" is; can it be mechanical, electromechanical, or only electronic? What functions must the machine be able to perform, and to what degree? By a wider and less technical definition, Babbage's Analytical Engine could be considered the first, since it had an instruction unit, a memory, automated input and output, and peripherals; but it was never built. George Stibitz built a binary adder in 1937 and a year later Konrad Zuse was the first to use binary in a calculator (his machine used discarded 35mm movie film as punched tape). Stibitz, Zuse, and Howard Aiken of Harvard University (Aiken developed the Harvard Mark I in collaboration with IBM) all independently developed *electromechanical* computers, using relays to represent numbers; Zuse's Z3 machine was "the world's first fully operational calculating machine with automatic control of its operations".[22] The claim of the first *electronic* digital computer, however, would go to the Atanasoff-Berry Computer (ABC), built between 1937 and 1942. Atanasoff's machine used electronic switches and capacitors to store numbers, although rotating drums of capacitors were still required for the memory.

Many of Atanasoff's ideas were later incorporated by John Mauchley, J. Presper Eckert, and others at the Moore School of Engineering, into the ENIAC (Electronic Numerical Integrator And Computer), built from 1943-1945. The ENIAC was the first *general-purpose* electronic digital computer; it used thousands of vacuum tubes, and could be reprogrammed for different problems by changing the cables plugged into a control panel. The only other machine operating at electronic speeds was the British Colossus, used for codebreaking during World War II, but it was not a general-purpose machine. The ENIAC is often cited as a "first" computer because it represented a quantum leap in computing power. As Paul E. Ceruzzi put it;

> The ENIAC computed about 500 times faster than any of the electromechanical computers, a difference of scale that made it an entirely different type of machine. With a relay machine it was always possible to measure computing power in terms of the number of human beings it could replace--that was the measure used in the published accounts of the Bell Labs machines and the Mark I. But the ENIAC was built precisely to tackle a job that by nature was beyond the capabilities of human--or electromechanical--computers.[23]

Yet the ENIAC, and all the machines before it, had to be programmed either by hand (through the replugging of cables) or by punched cards or tape. While the ENIAC was being constructed, the idea of a stored-program computer came about, although the question of who was the first to invent the concept remains unsettled. Plans for the EDVAC (Electronic Discrete Variable Computer) were started in 1944 by the ENIAC team, along with mathematician John von Neumann. The EDVAC was to be the first *stored-program* computer, but it was not finished until 1952. In January of 1948, the IBM Selective Sequence-Controlled Electronic Calculator (IBM SSEC) appeared, which had 13,500 vacuum tubes and 21,400 mechanical relays; so it was a stored program computer, but unlike the ENIAC, it was not fully electronic. Later that year, in June, the Manchester Mark I, at Britain's Manchester University, became "*the first fully electronic stored-program computer in operation.*"[24] And finally, a few months later in September of 1948, additions to ENIAC made it into a stored-program computer as well, although that had not been in the design of the original architecture.

With the invention of the stored program, all the major ideas and concepts behind the construction of the digital computer were in place. After the 1950s, further developments in computer technology were not measured so much by their speed, but rather by the types of technology or architecture employed, and these developments are grouped together into "generations" of computer technology. The ABC, ENIAC, Colossus and other early models are considered to be prototypes existing prior to the first generation, or a "zeroth generation"; the first generation starts not with prototypes, but with the first *commercially* produced computers, which used miniaturized vacuum tubes, had central processing units and were reliable enough to be commercially viable. By this definition, the first generation began in 1951; in February of that year, the first Ferranti Mark I (Ferranti Ltd.'s computer, based on the Manchester Mark I) was delivered to Manchester University; and in March the first UNIVAC (developed by Eckert and Mauchley) was delivered to the U. S. Census Bureau, becoming the first large-scale commercially-built computer.

The first four generations are each associated with a particular technology that they used (dates given are approximations): first generation (1951-1958)[25] computers relied on vacuum tubes; second generation (1958-1963) ones used transistors; third generation (1963-1971) ones had integrated circuits (ICs); and fourth generation (1971 to the late 1980s) computers had very large-scale integrated circuits (VSLIs). The uniqueness of the fifth generation is not so much one of hardware as it is architecture; rather than being "smaller, faster, cheaper", fifth generation computers are said to be "more intelligent".[26] But not everyone agrees with the ambitious claims made for the developing technology, and there is some debate as to what exactly constitutes the "fifth generation". And still less loosely defined is the "sixth generation" of computers occasionally mentioned in technological discussions, which also refer more to software architectures than hardware advances.

Throughout all of these developments, the bit is still represented by a presence or an absence, although the entity representing the bit has shrunk from punched holes and relays to ever decreasing groups of electrons or magnetic particles. The bit represents the greatest degree of abstraction possible in the representation of information, and whatever the information itself in turn represents. "Information" itself has become a purely formal term, and every operation information undergoes can be expressed as symbolic manipulation. The workings of the computer where the bits commute over circuits are no longer visible; their elements have grown so small that they can only be seen through electronic imaging devices, which are often themselves computer-controlled. The designing of computers is now so complex that it can only be done on a computer; and more and more of the design process is being automated.

The uses to which computers are put are also changing. The computer's original uses were military and industrial, due to the large amount of capital needed for its development (a problem that slowed and eventually defeated Charles Babbage), but the commercial availability of computer equipment, and especially the personal computer, have broadened its sphere of influence into everyday life. Since the 1960s, when computer art pioneers such as John Whitney and A. Michael Noll found new ways of using the computer in the creation of art, the computer has changed existing notions of art, some of which were already in flux from the beginning of the century.

Digital technology continues to develop today, and has become the basis for a number of fast-growing industries and commercial markets. Few technologies have so quickly been integrated into daily life and become so heavily depended on. The computer's effects are often subtle but wide-ranging, and recent enough that they have yet to be explored in full. In shaping art, media, and communications in general, digital technology has changed the nature of cognition itself, and the way in which human beings think of the world.

NOTES

1. In some languages, like Dioi, spoken in southern China, numbers standing alone are considered improper. See Ascher, Marcia, *Ethnomathematics: A Multicultural View of Mathematical Ideas*, pages 12-14.

2. Rotman, Brian, *Signifying Nothing: The Semiotics of Zero*, St. Martin's Press, New York, ©1987, page 1.
3. The information on Hariot, Bacon and Lobkowitz comes from Glaser, Anton, *History of Binary and Other Nondecimal Numeration*, Los Angeles, California: Tomash Publishers, ©1971, pages 11-30.
4. Ibid., page 35. Most of the rest of the letter appears in Glaser's book on pages 31-35.
5. Although Boole did not use binary digits, they are one of the clearest examples of how his operators work: the NOT operator takes a single input and returns its opposite (a 1 becomes a 0, and a 0 becomes a 1). The AND operator returns a 1 if both of its two inputs are ones, otherwise it returns a zero. The OR operator returns a 1 if one of its two inputs are ones, but not if they both are. These three functions are the basic building blocks for other more complex ones.
6. As William Aspray points out,

> Shannon was not the first to suggest this isomorphism. The idea had been suggested in the Russian literature in 1910 by Paul Ehrenfest and followed up in 1934 by V. I. S. Sestakov. It also appeared in a 1936 Japanese publication by Akira Nakasima and Masao Hanzawa. However, none of these received the wide attention of Shannon's paper, mainly because his paper was in English and presented a detailed account of the isomorphism in a way that highlighted its value to circuit design theory.

Aspray, William, editor, et al, *Computing Before Computers*, Ames, Iowa: Iowa State University Press, ©1990, page 117.
7. The quotes are from Shannon, Claude E., "A Symbolic Analysis of Relay and Switching Circuits", *Transactions of the AIEE* 57:713-723, December 1938; and Shannon, Claude E., "The Mathematical Theory of Information", *Bell Systems Technical Journal*, July 1948, page 380.
8. John Horgan explains Wheeler's idea in his article, "Quantum Philosophy" in *Scientific American*, July 1992, page 104.
9. Augarten, Stan, *Bit by Bit: An Illustrated History of Computers*, New York: Ticknor & Fields, ©1984, page 18-22.
10. Williams, Michael R., *A History of Computing Technology*, pages 172-173. The largest completed and working difference engine of Babbage's time was built by G. B. Grant; it was five feet high, eight feet long, weighed over a ton, and contained over 15,000 parts. During the early 1990s, Babbage's Difference Engine No, 2 was constructed at the British Science Museum, based on his plans, and it worked.
11. Augarten, Stan, *Bit by Bit: An Illustrated History of Computers*, page 74. For a detailed account of Babbage's life and his engines, see Anthony Hyman, *Charles Babbage: Pioneer of the Computer*, Princeton, New Jersey: Princeton University Press, ©1982.
12. Ord-Hume, Arthur W. J. G., *Pianola: The History of the Self-Playing Piano*, George Allen and Unwin, Publishers, London, ©1984, page 9.

13. Ibid., page 199.
14. Broudy, Eric, *The Book of Looms: A History of the Handloom from Ancient Times to the Present*, Brown University Press, Hanover and London: The University Press of New England, ©1979, page 134.
15. From *The Cyclopaedia of Useful Arts*, edited by Charles Tomlinson, published London and New York, 1862, Volume II, page 959. A full-page black and white reproduction of the portrait appears across from page 113 in Anthony Hyman, *Charles Babbage: Pioneer of the Computer*, Princeton, New Jersey: Princeton University Press, ©1982.
16. From the Guest Editorial "Telecommunications in China" by Ye Peida and Richard Skillen, in *IEEE Communications Magazine*, July 1993, volume 31, number 7, page 14.
17. Chappe's and Edelcrantz's optical telegraphs are described in "The First Data Networks", by Gerald J. Holzmann and Björn Pehrson, in *Scientific American*, January 1994, pages 124-129.
18. For a concise explanation of Huffman encoding, see Stix, Gary, "Encoding the "Neatness" of Ones and Zeroes", *Scientific American*, September, 1991, page 54.
19. Roblin, Jean, *The Reading Fingers: Life of Louis Braille 1809-1852*, translated from the French by Ruth G. Mandalian, New York: American Foundation for the Blind, ©1955, page 32.
20. Stibitz, George, as told to Evelyn Loveday, "The Relay Computers at Bell Labs", *Datamation 13*, April, 1967, page 35. This quote is also found in Augarten, Stan, *Bit by Bit: An Illustrated History of Computers*, page 100.
21. H. P. Babbage, *Babbage's Calculating Engines*, London: E. & F. N. Spoon, 1889, page 308. Reprinted by Tomash Publishers, Los Angeles, 1982.
22. Michael R. Williams, *A History of Computing Technology*, page 411.
23. Ceruzzi, Paul E., *Reckoners: The Prehistory of the Digital Computer, From Relays to the Stored Program Concept, 1935-1945*, Westport, Connecticut and London, England: Greenwood Press ©1983, page 105.
24. Augarten, Stan, *Bit by Bit: An Illustrated History of Computers*, page 148. The question of which machine qualified as "the first computer" was a controversial one precisely because the computer was still being defined, and so which one was considered the "first" depended on a definition which kept changing as the technology matured. Also, because the Manchester Mark I was developed in Great Britain, the national claim to the "first" computer was also at stake.
25. The first generation ends in 1957, 1958, or 1959, depending on the source consulted. Of course, these generations overlapped; years dividing them can indicate either the year when the technology began, or the year when it became the dominant technology. Authors are not always clear as to which division criterion they are using.
26. Bishop, Peter, *Fifth Generation Computers: Concepts, Implementations and Uses*, New York: Ellis Horwood Limited, Chichester, England, and Halsted Press: a Division of John Wiley & Sons, ©1986.

II.

Art

3.
The Work of Art in the Digital Age

Quantized art could only come about in a world of quantized time and space; with the advent of digital technology, the stage was set. From its origins in the laboratory to its appearance in the commercial world and the home, digital media have changed art as much as did the process of mechanical reproduction, which itself has been changed by digital media. One of the greatest of these changes is the additional mediation placed between artist and audience; for digitally-stored artwork, some medium, like pinned barrels, punched cards, or CD-ROM, is needed to "contain" it in its abstract, encoded form. This divide between "software" and "hardware" is found in musical scores and scripts of plays, which had to be performed in order to be experienced. Yet people could still read the script or score and get something out of it, or perform it themselves. Film, however, relied on a machine for its "performance"; one could look at frames of film without a projector, but most of the effect would be lost. Edison's wax cylinders for recording sound and phonograph records were even more reliant on machines, and could not be read directly by human beings. Electronically-stored and digitally-stored works of art also require machines to read them, continuing the division between software and hardware.

Walther Benjamin's famous essay of 1936, "The Work of Art in the Age of Mechanical Reproduction" looked at how mechanical reproduction changed the work of art, and digital technology brings with it still more changes. Benjamin's essay includes a number of overarching concerns: authenticity and the idea of the 'original', 'aura' and cult value, the fragmentation of the role of the actor, and changes in human apperception. Digital reproduction results in changes in each of these areas, raising new questions in many of them (some of which will be examined later in this chapter).

While mechanical reproduction attempts to reproduce a series of (nearly) identical objects, electronic reproduction reproduces an analog electronic signal, and digital reproduction involves the reproducing of a series of absences and presences used to store information. Thus overlap can occur; digital reproduction can occur mechanically (as in a punched card reading machine which punches new cards) or electronically, as in a computer. Electronic reproduction is sometimes digital, but need not be; radio and television transmissions are analog signals. The three forms of reproduction should not be seen as a series of movements or advancements replacing each other; rather, they are strands that developed and grew at differing rates, occasionally merging with each other. Even before computers, digital reproduction occurred, but only as a special case of mechanical reproduction, differing from other forms of it in only a conceptual sense. The combination of electronic and digital reproduction is what brought with it the 'digital age'. Neither mechanical nor electronic reproduction (in its analog form) could reproduce things perfectly, nor could digital reproduction while it was still yoked to mechanical means. Thus digital reproduction, today, has come to mean *electronic* digital reproduction, and it is principally this kind of reproduction to which I will be referring in this chapter.

Production

While many analog media can be manipulated directly (i.e., putting paint on a canvas, or dodging and burning a photographic print), the production of digital media requires the extra step of representing an artistic design or composition into machine-readable digital form, or requires the artist to produce it directly in digital form. In addition to preparing a work for audience reception, the machine must be considered as a stage of production situated between artist and artwork.

The earliest digital works of art (that is, *digitally stored* works of art) were the compositions on the pinned-barrel music machines and images woven on looms controlled by punched cards or paper rolls. Mosaics made from square tiles and hand-woven imagery had been around for centuries before these devices, but were not truly digital; although their design was based on a grid system, the information needed to reproduce them was not stored in digital form, though tiling and weaving were important precursors of digital art since they developed methods of design for translating images onto a grid.

The designing of images to be woven on looms using the Jacquard machine was a particularly premeditated activity. Designs had to be quantized onto grids, and then encoded into series of punch cards. The technical skills required for the process resulted in a series of books on the art, which often discussed aesthetics along with technical matters. Works like E. A. Posselt's *The Jacquard Machine Analyzed and Explained* and T. F. Bell's *Jacquard Weaving and Designing* covered the entire process, including the design of the imagery, which was to be done on graph paper. Posselt explains how to develop patterns that can repeat without breaks, along with a "squaring off" process for enlarging, reducing, or duplicating sketches. Following that are instructions, called "Outlining in Squares", for transferring a sketch to squared paper —a method of breaking the image into pixels, done by hand.[1] The "outlining in squares" system is essentially similar in concept to the image scan-conversion algorithms used by computer for converting imagery into digital form. Working with graph paper changed the method of design, and even the traditional artist's tools had to be adapted to this new system:

> The brush used by the designer must be clipped according to the size of the rectangles of the paper. It should cover the rectangle in warp direction at one sweep of the hand; hence each size of the squared paper requires a specially prepared brush for quick, good, and perfect work.[2]

Throughout the text, examples are given of patterns and their translation to grids; one figure shows concentric circles of different sizes and their representation in squares. Many of the images are highly detailed and surprisingly elegant, due to the long history of pattern design for weaving, as well as due to the Jacquard machine itself. Images made with the Jacquard machine could range from simple repeating patterns to finely detailed scenes taken from paintings; some, like the portrait of Jacquard described in the last chapter, had as many as 1000 threads per inch (in both warp and weft), a high resolution for a woven image even today. The creation of lighting effects and shading, through the use of processes similar to dithering, is also described and encouraged in Posselt's book and design instructions are given for different types of fabric.

Light and shading effects, for depicting round or dimensional forms, are also strongly encouraged in T. F. Bell's book of 1895, *Jacquard Weaving and Designing*. Bell's book goes into much greater detail (with numerous examples, some as large as 104 by 176 squares) and has chapter entitled "Designing and Draughting". Bell also makes a distinction between the design of an image and the drafting of it onto squared paper, showing the work to be done in two steps. Further mediation between artist and artwork was already occurring; artists had to learn new skills —the "squaring off" of designs— if they were to maintain control. The separation between design and execution made it possible for the designer to leave production to an assistant or a machine. Even today, needlepoint kits sold commercially usually contain designs supplied by the manufacturer.

Whereas imagery had to be quantized on graph paper before it could be stored digitally, music was already quantized, as it had been for some time, in the form of scales (diatonic, chromatic, modal, microtonal, etc.) represented by forms of notation (although the number of divisions varied, most scales are based on the quantization of frequency into a series of notes). As elaborated in the previous chapter, music was pinned onto barrels, an act as painstaking and precise as the "outlining in squares" described above, leaving little room for improvisation. Reproducing pianos, developed around 1904, were the first means of actually recording a performance in real time, and storing it in digital form; but here, too, the new technology changed the nature of production. First, the range of expression was severely limited. According to piano historian Arthur Ord-Hume;

> It is obvious that not all pianists were enamoured of the reproducing-piano system. There is a great story about Artur Schnabel which demonstrates this. On being asked if he would record for Duo-Art, he declined. Pressed by Aeolian with the advice that the Duo-Art offered sixteen shades of nuance, Schnabel replied with the news that in his playing he used seventeen shades.[3]

The interposition of the machine between the pianist's performance and its audience gave no guarantee as to what it would sound like at later playings of the recorded paper roll; while the original recording was made on a particular piano under studio conditions, player-pianos used to play the roll might be out of tune, or situated in acoustically dead rooms, resulting in an inferior performance.

And, as is always the case in modern recording studios, other hands were at work during and after the performance:

> This began with the mysterious operation of manual controls on a separate control box as far as Duo-Art and Welte rolls were concerned, so that the work of the pianist received its first interference from an outside hand as he was actually playing. Then came the eradication of wrong notes, the correction of imprecise phrasing, the insertion of missing notes and minor adjustments in *tempi*. The final master roll was then a mixture of recording pianist, roll editors and engineers.[4]

Thus in both graphic art and music recording, the interposition of a machine often resulted in an increased reliance on technicians who mediated between the artist and the final work of art, occasionally to the degree of a collaborator.

The only way the artist could avoid this mediation was to become a technician. Painters and sculptors had always used tools; it was now a matter of including digital technology. The new technology had inherent limitations, but so did the old; and it was from precisely these limitations that distinct styles and effects had emerged. Unlike Samuel Morse who gave up his career as a painter to work on the telegraph, the artist could combine art and technology and find what new forms could be produced with it. The acceptance of the machine in both daily life and art contributed to the Futurist and Constructivist movements of the 1910s that exalted and glorified the machine, using it as both medium and subject matter. As early as 1912, the Futurists planned moving sculptures driven by machine, in which the machine was incorporated into the body of the artwork itself. Both movements advocated the artist as a user of technology and recognized the machine as art, ideas that would slowly make their way into the artistic mainstream in time for the arrival of electronic digital technology.

After digital technology developed into the computer, artists soon found new uses for it. The first graphic images generated by an electronic machine were the "oscillons" or "electronic abstractions" made on an oscilloscope by mathematician Ben F. Laposky, in 1950. By 1960, John H. Whitney was already using a mechanical analog computer, made from anti-aircraft gun directors, to mechanically move animated typography and abstract designs; in 1966 he received a three-year research grant from IBM and began producing short films. During the 1960s, the line between artist and technician began to blur; artists like John Whitney, Lillian Schwartz, Peter Foldes, Ed Emshwiller, Stan VanDerBeek, and John Cage began to use the new technologies in their work, while scientists like A. Michael Noll, Ken Knowlton, Leon Harmon and Bela Julesz of Bell Labs and mathematicians George Nees and Frieder Nake began to use their technologies to create art. Collaborations often arose between the two sides; Knowlton assisted VanDerBeek and Schwartz on many of their films, IBM funded artists like John Whitney, and from 1965 on, computer art exhibitions began to appear in art museums.

The juncture, however, was somewhat premature: equipment malfunctioned during exhibitions; some of the work produced was mere novelty; artist and scientist were occasionally at odds; and exhibitions were often poorly done. The successful joining of art and technology, on a widespread basis, had to wait a few years until the technology was sufficiently developed enough to leave the laboratory and the technician's supervision, allowing the artist greater control. This finally came about during the 1970s and early 1980s when computer technology became user-friendly and inexpensive enough to become a commercial product.

The computer differed from other artists' tools; besides the synthesizers and paintbox programs that were used to create a work, the computer itself could be programmed to create a work. In a sense, artists had already been creating programs, to be carried out by other people if not machines; for example, Mozart had ways of composing using random throws of dice, and playwrights wrote scripts that were performed by others (the word "program" itself originally referred to a list of events, as might be found at a concert or play). Unlike a written play, however, a computer program is very precise concerning the means of execution, and a programmer can be more assured of getting a specific result.

The extended role of the computer as tool made necessary the distinction between *computer-assisted* art and *computer-created* art. In computer-assisted art, an application like a paintbox or music program is used in the creation of a work. Computer-created art begins with an algorithm written by the artist, which when run, produces a work, usually with some element of randomness involved so that different works can be produced with the same algorithm. This is especially true in certain forms of music, where algorithmic methods of composition existed hundreds of years before the computer. The algorithm, then, becomes the work of art; it is both *process* and *product*. It is the nature of the artist's methods distilled from the raw materials affected by those methods; it is the process *as* a product.

Some of the earliest and most interesting examples of the algorithm as art can be found in A. Michael Noll's work. The best known of these is his *Computer Composition With Lines* (1964), a computer-generated design of pseudorandom elements statistically based on painter Piet Mondrian's "Composition With Lines" (1917). According to Noll,

> Although Mondrian apparently placed the vertical and horizontal bars in his painting in a careful and orderly manner, the bars in the computer-generated picture were placed according to a pseudorandom number generator with statistics chosen to approximate the bar density, lengths, and widths in the Mondrian painting. Xerographic copies of the two pictures were presented, side by side, to 100 subjects with educations ranging from high school to postdoctoral; the subjects represented a reasonably good sampling of the population at a large scientific research laboratory. They were asked which picture they preferred and also which picture of the pair they thought was produced by Mondrian. Fifty-nine percent of the subjects preferred the computer-generated picture; only 28 percent were able to identify correctly the picture produced by Mondrian.[5]

The choice of Mondrian is, of course, one of the key reasons for the outcome of the experiment. Because of the limitations of the early technology, most early computer art tended to be abstract (usually involving random elements), and used purely conceptual designs. They were usually minimalist, geometric and linear, often involving symmetry and repetition.

By the mid-1970s and throughout the 1980s, computer graphics improved tremendously, and new kinds of imaging were possible: visualizations of complex mathematical functions, three-dimensional graphics, and fractal imagery no artist could have rendered by hand. Today, computers are found in almost every area of art, though with varying degrees of success, acceptance, and originality; machine reliance can become a crutch to the uninspired; in music, for example, "sampling" has occasionally taken the place of original creation.

The artist's participation in initiating the work of art and guiding its development is still necessary, if anything meaningful is to be produced, although as Noll's experiment demonstrates, abstraction erodes notions of what is "meaningful". In computer-controlled art and computer-created art (in which the art itself is generated by the computer according to an algorithm), the artist is often the one responsible for the algorithm used. Some algorithms, however, have become commercial products, complicating notions of authorship. One such program, *C. P. U. Bach* made by MicroProse Software Inc., claims that it:

> "...actually creates *original* classic music in the style of the legendary Baroque composer Johann Sebastian Bach. . . Concertos, fugues, minuets, chorales—the world's first true digital composer, C. P. U. Bach, handles all the demands of Baroque music, and yet still lets you adjust its performances to your own personal tastes. Those pieces you like best can be saved for playback anytime.[6]

While the ad bills the program as "the latest composition of award-winning software designer Sid Meier", the rest of the ad attempts to personify the program itself as the "artist". Should the music produced by the algorithm be considered a part of it, or entirely separate works of art? The creating of artwork by algorithm blurs the line between the artwork and the tool used to create it; perhaps even between the tool and the creative process itself. At the same time, it is often a break from traditional methods, displacing certain tools and skills. As Herbert Franke writes, the computer removes the need to acquire certain manual skills and techniques, into areas one could describe as cerebral.[7]

The shift from the manual to the cerebral coincides with the shift from the perceptual to the conceptual, present also in the shift from analog (dealing with the physical) to digital (dealing with the conceptual). Apart from the intervention of machine (and technician), digital technology's main contribution to the production of art is the extension of the degree to which the artist can use a tool in the creation of a work of art. How much of the decision-making process involved in production can be handed over to a machine? We might also ask how this affects the importance or abilities of the artist, and if the term "artist" remains unchanged. One could even go further and ask what constitutes a work of art, as far as the artist's participation is concerned, and where the hand of the artist ends. Should music composed by *C. P. U. Bach* one hundred years from now be attributed to the algorithm, the programmer who wrote the algorithm, or should it be attributed, in part, to J. S. Bach, after whose style it is statistically modeled? Or the user who comes up with the parameters while using the program? Or maybe the "music" that it creates isn't art at all, while the algorithm that creates it is—or is *that* really J. S. Bach's as well?

The constitution of the work of art —what is considered important in its makeup— directly affects the physical preservation of the work of art, as it helps to determine how the work is to be preserved, as well as the exhibition of the artwork, which also often depends on a machine.

Preservation, Restoration, and Exhibition

Traditionally, most works of art have been physical objects, subject to the forces of nature which cause them to age and decay. Preservation attempts to lengthen the life of the art object for as long as possible, but all that one can really do is slow down the process of decay; paintings darken, film fades, and sculptures crumble. The digital work of art, however, is stored as information; and lacking physicality, it does not decay. The *medium it is stored in* —compact discs, magnetic tape, plastic cartridges— will age and decay, and dropouts can occur, obscuring the electrical charges on the magnetic tape or disc, but the artwork itself (or rather, the presences and absences of which it is made) can be transferred to another storage device without any loss. Analog forms of art are more inseparably bound to the physical media that make them up, and so they decay in analog fashion —gradually, like a darkening painting. Digital forms decay digitally; the 'zeroes' and 'ones' continue to be readable until a discrete threshold is passed in the decay of the physical medium. Once the zeros and ones are no longer readable, no trace of them remains, and the signal is lost.

The fixity possible with digital preservation has resulted in the digitization of art collections and the creation of digital archives. Digital preservation differs from traditional art conservation processes, which attempt to preserve the physical object along with the image it contains. For paintings, photographs, and other physical media, digital archives can be no more than records; lacking physicality, they cannot pretend to replace the objects they document. As yet, no worldwide standard exists for the digitizing of art. Digitized artwork will always carry the mark of the time of its digitization with it; the *Mona Lisa*, for example, if digitized today, would be stored with all the blemishes and cracks it presently contains. As a digital work it would remain frozen as it was at the time of its digitization, and would bear the marks of the methods used at that time; the time of creation of the work of art and the time of its digitization would both play a part in the appearance of a digitally-stored work of art. Works best suited for digital preservation and storage are those which are produced in digital media. When analog works are digitized and digital records are made of them, some works of art will be altered more than others, meaning that some are better suited for digital preservation than others (see chapter four for some the cultural biases inherent in the process of digitization).

Using Nelson Goodman's terminology, we can succinctly describe two general effects occurring during digitization: the conversion of one-stage works into two-stage works, and the conversion of autographic works into allographic ones.[8] Two-stage works are those whose production can be divided into more than one step, giving them the potential to be divided amongst two or more people; a musician can perform another person's score, a photographer can make prints from someone else's negatives, and so on; a painting, on the other hand, involves no intermediary stages and is thus considered as a one-stage work of art. All digital images, then, are two-stage works, no matter what their origin is; digitization means there is more than one stage in their production.

In *The Reconfigured Eye*, William J. Mitchell describes digital images as "two-stage, allographic, mechanically instantiated works." According to Mitchell;

> Autographic works such as paintings or videotapes consist of analog information: they cannot be copied exactly, and repeated copying always introduces noise and degradation. But the specification of an allographic work consists of digital information: one copy (of a musical score, of the script of a play, of an image file) is as good as another... Allographic works can be instantiated limitlessly (but the concept of instantiation does not apply to autographic works—they are unique): a musical work is instantiated in a performance that faithfully follows the score, a play is instantiated in a performance that faithfully follows the script, and a digital image is instantiated in a display or print that faithfully follows the tones or colors specified in the image file.[9]

The conversion of an autographic work into an allographic one means a translation into notation; digital notation in the case of digitized works. Thus, it is not enough to preserve the digital files representing the artwork; the hardware and software which reads the notation and translates the files into imagery must also be preserved. Digitally-stored artworks requiring extinct machines or ones no longer functioning properly are more or less lost, as are those requiring programs written in convoluted "spaghetti code" or now-defunct programming languages and file formats. Even if all of their ones and zeroes remain intact, the work of art is indecipherable without the context provided by the correct hardware and software. As allographic works, digital artworks are dependent on interpretation, and as digitally abstracted ones, their interpretation is almost wholly dependent on machines.

Autographic works have a uniqueness that is lost when they are translated into allographic (and digital) form; but this is usually felt to be a small price to pay for the fixity that the work is given. The immortality available in digital form can have an especially strong draw to artists, and appeared even before the computer technology was available. In 1960, Peter Kubelka produced the very first digital film, *Arnulf Rainer.* The film was a few minutes in length and minimalist in form and content; the images were either all-black or all-white frames, the soundtrack either bursts of white noise or silence. The simplified sound and visuals meant that the sequences of both could be written down and easily reproduced. Enthusiastic about the fixity available in digital form, Kubelka wrote:

> With this film I have done something which will survive the whole film history because it is repeatable by anyone. It is written down in a script, it is beyond decay. It can be made exactly after my script without any faults. Someday I will put the script in stone, granite, then it will last 200,000 years, if it is not destroyed. You see the normal age of black & white film is 50 years. Now it is only 15 years, since the chemicals and materials are much worse than they were in 1905. A color film rots in about 10 years and then the colors fade and decay. In the entire world there is no way to preserve a color film. But *Arnulf Rainer* will last forever.[10]

At two bits per frame (one for picture, one for sound), *Arnulf Rainer* could be stored in only two or three kilobytes, making it one of the most compactly stored digital films as well.

The most ubiquitous use of digital preservation in art, however, has been in the sound recording industry, following the appearance of the compact disc. Better preservation means fewer replacements need be sold, and even the market for used CDs grew quickly. But the selling of used CDs reduces the earnings of record companies, distributors, publishing companies, recording artists, songwriters, and producers; it became one of the most disturbing issues in the sound recording industry.[11] Thus, better forms of preservation —in the hands of consumers— can cut into sales and profits, and companies may not always see it as in their best interests (excepting, of course, profits made in the changeover from records to CDs, in which old albums were resold on CD).

Finally, when one speaks of preservation, one must ask what is —or needs to be— preserved. As mentioned above, digital artwork stored as it is in abstract form, is very context-dependent. The context surrounding the digital work of art —in this case, a decoding algorithm that can make sense of it— must be preserved along with the digital file itself; if the decoding algorithm is lost, all files depending upon it become meaningless. This drawback of digital preservation is already beginning to appear. As Michael Gruber writes in *Scientific American*:

> NASA, for example, has huge quantities of information sent back from space missions in the late 1960s stored on deteriorating magnetic tape, in formats designed for computers that were obsolete 20 years ago. NASA didn't have the funds to transfer the data before the machines became junk. has "thousands of terabits" of data on aging media that will probably never be updated because it would take a century to do it. The archival tapes of Doug Engelbart's Augment project- an important part of the history of computing- are decaying in a St. Louis warehouse.
>
> "The 'aging of the archives' issue isn't trivial" says desktop publisher Ari Davidow. "We're thinking of CD-ROM as a semi-permanent medium, but it isn't. We already have PageMaker files that are useless."[12]

Digital preservation is insufficient protection against the ravages of history, which destroys the contexts of data files, relegating them to obscurity as it does the historical artifacts whose meaning is long lost, or runes written in some long-forgotten tongue.

Nor does digital preservation preserve the history normally embedded in an object. As Walther Benjamin pointed out, part of the "aura" of the work of art comes from "the history to which it was subject throughout the time of its existence. This includes the changes which it may have suffered in physical condition over the years as well as changes in ownership."[13] Thus restoration can, in a sense, be an effacing of history.

Restoration is perhaps one of the controversial forms of preservation, and although it can remove historical effects, it can also become a historical corrective. For example, during the mid-1980s restoration of the Sistine Chapel ceiling, the removal of accumulated dust and debris revealed bright colors underneath, suggesting Michaelangelo was one of the major colorists of his day. Leonardo DaVinci's fresco *The Last*

Supper is also undergoing restoration; Italian restorer Pinin Brambilla has removed dirt, overpainting, and substances applied during previous restorations. According to a 1988 article in *Art News*,

> Brambilla has been courageous, removing St. Simon's long beard, leaving him with the short goatee Leonardo gave him. The profiled face of Judas is now a three-quarter view. In front of St. Matthew, Brambilla has uncovered a slice of lemon on a plate. The glasses on the table can now be seen to have delicate gold rims, and the silver dishes reflect the colors of the apostles' robes. A loaf of bread that had appeared to be untouched can now be seen to be broken.[14]

Two and a half years later another *Art News* article would report,

> Brambilla has lifted enough repaintings to have significantly changed the tone of the painting. "The figures are much more animated than they were before," says Marani. "Even though there is less color on the wall, there is more spatial volume. We are discovering a painting of extraordinary beauty. It is like seeing Leonardo for the first time.[15]

Restoration of an artwork, then, can also change or restore the history behind it; as these examples show, it can affect the reputation of a work or an artist. These analyses, and new ones, can now be done on a computer. One procedure, according to Lillian Schwartz,

> involves using a scanning microdensitometer, which will scan and digitize a transparency of an image into its RGB components. Infringing colors, such as the black of the underlying charcoal sketch or a color portion that was painted over, can be enhanced. A sample of yellowed varnish can be analyzed with an optical spectrometer, which measures the transmission of red, green, and blue through the yellow. Once the percentage of loss is assessed, the colors can be corrected in the digitized image. A smoothing technique is used to cover up age cracks. . . . A photograph or slide of the painting or fresco is scanned and digitized so that a variety of calculations can be made. For example, the image can be mapped to show the areas of least and most craquelure. The greater the disparity, the more likely that the least-cracked portion was restored. A palette of that area can then be compared with a palette of the older area to uncover the restorer's deviation from the latter. The image can subsequently be enhanced by removing the craquelure and the yellow from the varnish and by applying the older palette to the restored portion. The computer can also use the information on the palette to reconstruct heavily damaged surface areas where the underlying sketch is available through infrared or X-ray techniques. Missing edges and other aspects of object definition can be discovered by means of common procedures such as

color fill, palette manipulation, and a switch from color to black, grays, and white.[16]

Digital restoration processes are also being used for restoring and replacing old film prints; in 1992, Disney's *Snow White and the Seven Dwarfs* was the first feature film to be restored using the high resolution Cineon digital film system developed by Eastman Kodak Company. According to Kodak, the project required the restoration of some 119,550 frames of 35mm film, each of which was scanned into digital format and were stored at 40 megabytes per frame. During the restoration,

> A network of image computing workstations was used to restore the digital pictures. Flaws embedded in the original film were corrected using innovative computer software. For example, fine dust on the glass platen of the animation camera stands used in 1937 was frequently photographed onto the film.
>
> Workstation operators were able to eliminate those flaws using a revolutionary "dust-busting" workstation program developed by Kodak and refined at Cinesite. Other imaging workstations were used to correct colors, "paint out" scratches and repair other damage which marred the film through normal handling and aging over a period of time.[17]

The removal of cel dust and correcting of faded colors may remove a certain element of history from the film, but like the frescoes cited above, it preserves the artists' original work. Digital preservation, of course, can only be done on digital files; is the image stored as a digital file the work itself, or only a representation of the work? For that matter, does the "work" always have to *be* an object? In the case of painting, and other autographic works, a digital file can only be a representation; but for allographic works like film, a computer file and a camera negative can be used to produce prints which are indistinguishable from each other. Still, while the digital reprints of *Snow White* can fulfill the same role as the analog ones, no digital version of *The Last Supper* could ever come close to replacing the fresco. Preservation and restoration effect not only the appearance of an artwork but its history as well, and are factors influencing an artwork's exhibition.

When a digital work of art is exhibited, it must appear in analog form, since all types of output devices are physical in nature. When this results in the production of an object, like a printout, a sculpture, or photograph, the object reenters the realm of the traditional physical art object, existing in tangible form, accumulating a history, and prone to decay. Most digital artwork, however, remains a series of images and/or sounds, to be played in sequence, requiring a machine to interpret and play them. Whereas human beings were traditionally the performers of an artwork (as actors or musicians), machines have become the performers of digitally-stored artwork, and their interpretation or performance is (hopefully) standardized with as little variation as is technically possible.

This brings us to the audience, the endpoint of the journey taken by the work of art, and the relatively recent redefinition of the audience as "users" of a work of art. The term "user" implies a shift to a utilitarian mode; instead of being merely "exhibited", some works (those referred to as "interactive") are "used". "Use" implies a purpose or function, or a mode of interaction. In another sense, the word also signals a shift in the activity occurring, just as the word "computer" once referred to a person doing calculation. Once the machine came to be known as the "computer", the person operating it became the "user".

The term is usually applied to home usage. Digital media expand the possibilities of home exhibition, but come with requirements; machines and some technical ability are needed, and reliance on high-tech companies is inevitable. New reading abilities —dealing with random access, menus, nonlinear form, and authoring possibilities— are needed, and more viewing options are available. Users must adjust to the increasingly complex visual field with simultaneous points of interest, although there are also more possibilities for repeated viewings and close analysis.

Digital media for the home can encourage more exhibition of artwork, but what gets lost in translation? When museum collections are translated into the home, for example, via the Voyager Company's laserdiscs of the Louvre collection, much of the viewing experience is lost, due to limitations in resolution, color, size, context, etc., just as full-motion full-resolution video loses a great deal as a low resolution QuickTime clip. While on one hand it can be argued that people can see works of art they wouldn't otherwise get to see, the images on-screen cannot completely represent the works they claim to; just as there is a difference between listening to a compact disc and a live orchestral performance. The user's experience can be no better than the reproduction allows.

Machine mediation effects the exhibition of the digital work of art in the analog realm as well. Performance is restricted by the machine; how well it is working, the availability of electrical outlets (or batteries or generators), and so on. Machines of some kind are almost always required for the digital artwork to be experienced. In some sense, then, the limited appearance of the work of art, which is "gone" when the machine is turned off, encourages the return of aura and cult value that Benjamin also wrote about.

The Return of Aura and Cult Value

If, as Walther Benjamin writes, "that which withers in the age of mechanical reproduction is the aura of the work of art",[19] then that which withers in the age of digital reproduction is the *physicality* of the work of art. The digital artwork, stored electronically, is invisible; it can only be experienced when deliberately put on display. So, without physicality, can a digital artwork produce the "aura" of the traditional work of art?

Part of the answer to such a question involves the idea of the reproduction versus the original. Digital reproduction is the only completely form of reproduction which can be completely accurate. Reproductions produced electronically, and especially mechanically, are subject to slight variations. Generational loss in analog media points backward to a "zero" generation, from which the copies were made; the "original" signal. In film-based photography, there is an original negative from which prints are made. These points of origination are neither completely autographic nor allographic, and they demonstrate how Benjamin's belief in the disappearance of the original was premature. Even in digital media, there can be an "original" data file from which other copies are made; but once made they are indistinguishable from each other. What is important to Benjamin, however, is the "unique existence of the work of art" that includes "the history to which it was subject throughout the time of its existence. . . changes in physical condition over the years as well as the various changes in ownership."[20] But is this kind of existence, and physicality itself, a requisite condition for cult value, and a necessary ingredient of the "aura"? Benjamin describes the cult value surrounding certain objects, writing,

> Today the cult value would seem to demand that the work of art remain hidden. Certain statues of gods are accessible only to priests in the cella; certain Madonnas remain covered nearly all year round; certain sculptures on medieval cathedrals are invisible to the spectator on ground level. With the emancipation of the various art practices from ritual go increasing opportunities for the exhibition of their products."[21]

Cult value and exhibition value are closely linked; exhibition is necessary for cult value to develop, however, it must be limited so as to increase the anticipation of the object's appearance. Likewise, the digital work of art remains hidden, stored in abstract form, until summoned up by the user. This conjuring can only be done with the aid of a machine which brings up the imagery and sound from strings of invisible bits; a reappearing act which we have grown so accustomed to it no longer seems like prestidigitation. Sometimes passwords are even required before the work can appear, like magic words spoken during a conjuring act.

The machine, then, becomes like the veil or grotto; it keeps and safeguards the work of art, it is the window where the artwork can be seen by the user, until it vanishes again, when the machine is shut off (film and television function in a similar way). In some cases, the machine and its physicality can take on some of the cult value that the artwork has shed; from the ENIAC to laptops and pocket computers, there has been a growing fetish surrounding technological objects, beginning with the mystery and awe surrounding the room-sized computing machines of the immediate post-war period. The use of the computer, in which a person sits rigidly in front of it, sometimes for hours, also has a sense of ritual or even devotion about it. Like the ritualistic audience participation at a screening of *The Rocky Horror Picture Show* (1975) —a film with "cult" value if ever there was one— there can exist a sense of cult value and ritual around a reproduction. Film, television, video games, and some CD-ROMs like *Myst*, are not so much "objects" as they are experiences (or "events"), and often quite ritual-laden ones. But unlike the religious art that Benjamin writes about, these objects are the center of ritualistic practices; they do not aid in the performance of ritual, they are the reason for the ritual.

Dependent on, but not inseparable from, the physical objects (CD, magnetic disc or tape, etc.) in which it is recorded, the digital artwork is able to travel anywhere an electronic signal can go (unlike a physical object). Like the mechanical reproduction that Benjamin celebrated for being able to go places the original could not, "to meet the beholder halfway", electronic and digital artwork can be transmitted to a large number of receivers simultaneously, to wherever the hardware exists to receive it. Internet information, television, and radio transmissions have an ephemeral quality; after virtually instantaneous consumption by millions of people, they vanish without a trace. Only a small fraction is ever recorded by the audience; the artwork is reproduced and consumed without having become a physical object. Radio and television programs, internet newsgroups, and websites are received by millions (and often have cults surrounding them) and the sense of simultaneity they provide adds to the feeling of a shared experience.

Closely associated with cult value is the notion of the "aura" surrounding a work of art. Benjamin defines the aura as "the unique phenomenon of a distance, however close it may be."[22] No matter how physically close the work of art is, there is always a "distance" felt by the observer; as Paul Mattick, Jr. observes, "Object and viewer are not connected in the spatiotemporal framework of action; instead, the object appears as outside of time, a given, unalterable."[23] Since the digital work of art as a conceptual construct is not a physical object, there can be no way of measuring a physical distance from it; we can not approach "nearer" to what is ephemeral. Likewise, one cannot physically enter cyberspace; one can only be represented there. Even the sense of perceptual distance differs from that of the physical realm; an interesting example of this kind of "distance" can be found in the Mandelbrot set. The Mandelbrot set, among the most complex and detailed objects in mathematics, is a pattern resulting from the iteration of complex numbers, which are plotted on a grid.[24] Because it can be calculated for a grid of any resolution, one can "zoom in" indefinitely on any area of the Mandelbrot set, looking at its infinite levels of detail; and one will always be too "far away" to see everything—"distance" becomes a matter of *resolution.*

Benjamin felt the aura of the work of art was withering due to reproduction, but the notion of the "aura" has been reexamined and critiqued quite often in recent years. In his article "Recasting Benjamin's Aura", Patrick Frank sees aura as based on cultural authority and consensus, writing;

> If we see aura as a by-product of cultural authority rather than as genius, we will see that mechanical reproduction need not take away from the aura, but today helps to create it. This seems to be especially true of contemporary art, where consensus is still being formed. The role played by magazines in granting authority or "importance" to artists or styles is a widely accepted fact. Most works of art have little or no aura when created, but reproduction imparts to them a certain legitimacy that lends aura. This is as true of simulationist art as it is of any other kind. To try to create art without aura in the age of mass media is, to say the least, extremely difficult.[25]

Although Frank's use of "aura" is more broad than Benjamin's, his definition accounts for the aura of the digital artwork, and other signal-based artwork lacking physicality.

Digital reproduction, and reproduction in general, has also displaced aura onto its creator (the cult of the author or star performer) or its exchange value (millions of dollars, for some paintings), both of which are enhanced by publicity. The aura of the artist and performer, or even the exchange value of the material object, can be enhanced through reproduction even when the aura itself cannot be reproduced. Digitally reproduced works of art can maintain an aura about them, but the aura, and its attendant cult value, will always involve some connection back to the physical world.

Links Back to the Physical

On-screen train tracks converge at a distant vanishing point. Far away down the tracks, an old-fashioned locomotive turns a corner and begins heading straight for the audience. The image grows larger, gradually eclipsing the rest of the screen, its screaming whistle and chugging sound slowly building into a deafening roar. After moments of tense anticipation, the train fills the screen; the audience members instinctively duck and turn away —and then chuckle at their reaction. Except for the sound, the scene might have been an 1895 audience in France at a screening of the Lumiére brothers' film *Arrival of a Train at the Station at La Ciotat*, which reportedly caused audiences to run from the theater. The scene described above actually occurred at a screening of the IMAX film *To Fly* that I attended in 1981, 86 years after Lumiére's first screenings. The improved technology, the addition of sound and color, and the image size and clarity allowed the film to successfully illicit a visceral response in its audience even though they had been watching films for generations. Despite their often ephemeral quality, film and other media have always been able to create a visceral response in their spectators, ranging from the screams and beating hearts of the horror film to the emotionally wrenching melodrama (or network news) to the retinal fatigue of a flicker film.

Messages sent by electronic media differ from their predecessors in their lack of physicality. The analog signal, in its purest form, is a continuous stream of energy; when the stream is interrupted or frozen, as when it is stored on magnetic tape, it cannot be reconstituted *exactly*; some degradation, no matter how small, is always present after a transfer. Analog signals are continuous and represent their information with physically-measurable variations of some medium. They are more dependent on the physical media they inhabit than are digital bits, which represent their information more robustly through mere presence or absence; they exist conceptually instead of materially (although they rely on the material world for expression). But despite this detachment from physicality, digital artwork relies on the physical world, and has a number of links back to it.

The digital work of art always has two links to the physical world; through the medium in which the bits comprising the artwork are stored, and the system which reads, decodes, and translates the works into an analog form perceptible to the user. The digital "ghost in the machine" is powerless to operate without the machine.

Unique, autographic works of art, and even mechanically reproduced art such as prints or lithographs, engravings, molds, photographic prints, etc., exist as physical objects that have value in and of themselves, a value that is nontransferable to other objects (in some cases objects may even be part of a "limited edition" series, each with an assigned number that attempts to restore some of the object's 'uniqueness'). In any event, the artwork's physical composition is an inseparable part of it, and it is complete by itself, ready to be experienced without any further hardware or mediation necessary.

In contrast, the digital work of art cannot exist all by itself. Compact discs, magnetic discs and tapes, player piano rolls, video game cartridges, and even sheets of graph paper can all be considered as the "containers" which "hold" the series of bits comprising the digital artwork. Whereas the canvas and the paint upon it are inseparable from the painting as a work of art, and the material used in a sculpture inseparable from the sculpture as art, the "containers" inhabited by the digital work of art are not a part of the artwork, and do not share its value. The digital artwork is *dependent* on its container, but not on any *specific* container, and it is separable from it; a piece of music can be stored on a compact disc, an audio cassette, or a video tape, and conversely, the same cassette can have different pieces of music recorded on it. The value resides in the work of art and very little in its container (as is apparent in the difference in price between blank video tapes and ones containing commercial works), and this value is transferable, duplicatable, and erasable.

The data container's importance, however, should not be underestimated, for it gives the work an important physical manifestation. The container acts as the "location" of the work of art, it is the object which is made in multiples and sold in stores, it fulfills all the roles of the traditional commercial product, and thus serves to very successfully integrate digital artwork into preexisting commercial systems of exchange in which physical objects are bought and sold.

The role of commercial object performed by the container may decrease as new forms of commercial transactions develop to accommodate purely digital exchanges. Programs and data are bought and sold over electronic networks, where payments are made through electronic funds transfers. Since the containers are not part of the intrinsic value of the work and are separable from it, it is inevitable that industry will attempt to discard them once the marketplace had evolved to the point where it can do without them. But the new electronic systems are still not entirely in place, and systems of protection for digital products, like encryption and copy-protection, are still being improved upon. (The link back to the physical is sometimes used to safeguard programs which can be copied too easily. Rather than building everything into a particular program, which would be complete when copied, some game companies have a booklet with rules, passwords, and other information needed in order to play the game; at the start of the game, these passwords or secret codes are needed before play can begin, preventing pirated copies from being complete.)

Even when the data container is little more than a storage device, the machine that decodes and displays the data will still remain an essential link to the physical world. As the digital data depends on its container, so too does it depend on the "player" or machine (CD player, computer, game console) in which the container is inserted to be written on or read. The machine and the container share the same interchangeability as the work of art and the container; for example, as long as their formats match, a tape can be played in any player, and the player can play any tape. As the physical hardware of the machine controls the physical medium of the container within it (spinning a disc or winding a tape), the decoding algorithm within the machine (in the form of a program) processes the data in the container. The decoding algorithm is also dependent on the machine, as the work of art is on its container; the algorithm needs a machine in order to operate, but it does not need any specific machine (as long as the machine and algorithm are compatible). One difference, however between the work of art and the decoding algorithm, is that the decoding algorithm is often hardwired into the machine, and not completely separable from it (in a physical sense). Similar relationships can also exist between purely digital constructs; files are to applications what software is to hardware; an image file will usually need certain applications to be opened and used, just as the application needs hardware in order to run.

As the role of the container wanes, digital artwork itself is beginning to imitate the object-oriented commercial marketplace in which the container takes part. The term "object-oriented" is used in computer graphics where "object" is used as a metaphor for a collection of data which constitutes a mathematically-defined entity. Object-oriented computer graphics store an "object" as a set of instructions which describe the object mathematically, as a series of lines, curves, planes, and primitives (cubes, cylinders, spheres, and other polyhedra). These "objects" can be then moved, rotated, and manipulated as if they were single entities or units.

Besides being a useful metaphor for programming, object-oriented graphics also allow datasets to be thought of as discrete entities which can be bought and sold like physical objects. A good example of how the electronic marketplace can imitate the physical one, are the "Cyberprops" catalog from 3NAME3D or the "Dataset Catalogs" available from Viewpoint Datalabs. Just as in a catalog for physical objects, the objects are grouped into categories according to what they represent, a satisfaction guarantee given in the front of the catalog. Throughout, objects are shown as wireframe images, with the number of vertices and polygons given for each as a measure of their polygonal resolution. Varying resolutions are available for some objects, and files can be translated into "over 50 formats". Here is sampling of some of the datasets from Viewpoint Datalabs:[26]

Dataset	No. of Polygons	Price
Golf Ball	54530	$595
Cellular Phone	5624	$395
Shopping Cart	6938	$295
Ear w/Interior	33261	$595
Toyota Corolla 1991	4654	$395
Artificial Heart	36129	$349
Woman Pregnant	22327	$595
Beethoven	10791	$395
Buddha	37074	$695
Buffalo	8501	$349
Walleye	8434	$349
Toaster w/Internal Parts	21002	$395
Termite	62999	$695
MiG-29C	4132	$549
St. Basil's Cathedral	42536	$695
Pizza in a Box	441	$ 49

A quantized landscape of the United States is also available, in one degree by one degree quadrants each containing approximately 14,000 polygons. The catalog features mainly animals and vehicles, including dinosaurs and over 200 autos by company, model, and year. Military vehicles, including aircraft, helicopters, missiles, tanks, and watercraft figure prominently and are even listed by model and make. The contents make it clear that scientists, lawyers, the military, the entertainment industry, and ad agencies are the main consumers for whom the catalog is designed.[27]

Used in television commercials, these datasets stand in for the products being advertised; in a sense, Viewpoint DataLabs' "products" replace the actual physical ones in the ads, resulting in what could be considered false advertising. In this sense, computer generated "objects" not only mimic the role of physical ones, but even replace them in visual representations.

Besides standing in for commercial products, these "dataset-objects" could potentially be designed as representations of traditional works of art; not only the low resolution ones already available, but high resolution versions which could be used by galleries, for "virtual" shows. In any event, digital artwork enhances the aura of physical artwork by revealing its physicality and its inseparability from the object, which can no longer be taken for granted.

The rise in 'data-based' art lacking physicality has appeared and grown, perhaps not coincidentally, with a rise in the popularity of antiques and collectibles, whose physicality plays a key role in their value. This rise is also apparent in the changing meaning of "antique". As George Savage notes in his *Dictionary of Antiques*:

> The meaning of this word has changed very considerably during the last century or so. Originally it meant an object, usually classifiable as a work of art, which had been made in Greek or Roman times. Increasingly 'antique' has been taken to mean almost anything made before a certain date. Until recently the accepted date was 1830, because this was the year laid down by the U.S. Government as the limit for importation free of duty for many categories (furniture, porcelain, and so forth) which did not fall within the Customs' definition of works of art, and would otherwise have suffered the imposition of import duty. This date has now been altered to take in anything made one hundred years before the date of import, thus bringing the U.S. into line with most other countries admitting antiquities free of duty. . . . It is not possible to divide antiques and works of art satisfactorily, since the division in many cases is strictly one which has been made in modern times.[28]

Thus as mechanical reproduction brought with it a greater awareness of the "aura" or history imbedded in certain works of art, it also blurred the line between works of art and objects having a long history, adding to a growing interest in "antiques". Forgeries exist in both art and antiques; and "antique" can also be used as a verb, meaning to give an antique appearance to something by artificial means. Today, antiques are often displayed and as meticulously cared for as art; reproductions have made them valuable as "originals".

Although Benjamin considered the photograph to be a mechanical reproduction whose aura had withered, the collecting and auctioning of old photographs has become as widespread and lucrative as antique collecting. Today, photograph are collected by individuals as well as art galleries, auction houses, archives, and other institutions. Despite what Benjamin thought, one *can* ask for an "authentic" photographic print—

for example, the prints sold at auctions are often ones printed by the photographers themselves (Ansel Adams, Lazlo Moholy-Nagy, etc.) and are thus considered to be printed the way the artist desired them to be; they were physically manipulated by the artist in the same manner as a painting or sculpture. In 1990, 35 original Moholy-Nagy prints were purchased by Japanese buyers for $1 million, indicating how "authenticity" can reside in a photograph, which has a physical history as does an antique.

Mechanically-reproduced works of art are usually mass-produced as well, and while this can increase an object's familiarity, due to its widespread appearances, it also reduces its value as a rarity. Over time, however, the reduction of the number of circulating or extant copies (or complete sets, as in Depression Glass or baseball card collecting), can increase the rarity of an object, raising its value. This creates an interesting dynamic; an object is considered common initially, its value small, and does not appear important enough to save. If few people save the object, it becomes rare and hard to find, and once this is recognized, it can become a collectible and begin increasing in value.

Manufacturers, however, cannot afford to wait for the ravages of time and the changing tastes of a populace to turn their mechanically-reproduced items into collectibles; thus rarity must be built into the object itself from its very inception. Although one could build extremely fragile objects destined to become rare due to deliberate vulnerability, the answer often lies in a compromise between mass-production and singularly-produced autographic works; the concept of the "limited edition" collectible. These collectibles usually include some seal of authenticity on the object itself, and a serial number, often with some indication of how many of the objects exist. Some promise is usually made to insure that production will be limited; for example, in the production of Enesco's "Precious Moments" figurines, the mold is broken after the limited production run is finished, usually resulting in an immediate jump in price for the item in question. The serial number, which attempts to restore uniqueness to the mass-produced object, is where the value resides; if this is removed or lost, the object can lose much of its value and "authenticity".[29]

Unable to physically age and often easily reproduced, digital artwork cannot become an antique or collectible in the same sense as physical objects can, nor does it appear in limited editions. Only hardware can acquire such status, and some older game consoles and cartridges are already increasing in their value to collectors. According to a 1994 issue of *Wired*, the auction house Christie's put up for sale, "an assortment of computer games on the block, in the hopes, presumably, of creating a new sort of market for collectibles. Christie's will start with the more than 1,000 "classic" games (*Zork* and its ilk) gathered in researching a book called *The PC Games Bible*."[30] For some years now, Atari 2600 cartridges have also become collectible, with rare and hard-to-find cartridges going for higher and higher prices, and museums are getting involved in the production of traveling exhibits devoted to video games. "Hot Circuits: A Video Arcade" opened at the American Museum of the Moving Image in Astoria, New York museum in 1989, and since June of 1997, the large-scale exhibition "Videotopia" has been touring museums around the country.

In both game consoles and cartridges, programs are hardwired and fixed in the game cartridge itself; but most programs produced today are transferable and erasable, and not inseparably yoked to some physical object. Computer-generated "objects" and digital models have increasingly been able to emulate physical models in dynamics, and most of all in appearance. This has made them increasingly attractive to film, television, and other imaging industries, who see them as potentially becoming cheaper and more flexible alternatives to physical models. The replacement of physical models with digital ones, however, is not unproblematic, even where only the image is taken into account.

Digital Models vs. Physical Models

Toy Story (1995) was the first feature-length film to be completely computer-generated; all objects, scenery, lighting, and so on, existed only as digital models in a computer. While the film attained a remarkable degree of photorealism and realistic movement, it was still unmistakably stylized —no one would mistake one of *Toy Story*'s human characters for flesh and blood actors. In general, digital special effects have shown themselves to have great potential, but predictions as to the demise of physical model-based effects are premature. Technological breakthroughs inevitably increase the computer's abilities and role, but physically-based effects will still have their niche. Implications of the shift from physical to digital models extend beyond the film itself to the audience's reception of the film, the discourse surrounding the making and marketing of a film, and the choice and handling of subject matter in the filmmaking process. The roots of some of these implications can be found in the basic assumptions of digital modeling process.

The mathematical basis of digital technology, and particularly computer graphics, is unavoidable; any digital representation of an object will require a mathematically expressible representation of that object. What matters, then, is the degree to which mathematics is subsumed into higher functions that simulate objects and their interactions, how quickly and easily they can be manipulated by the artist, and the degree of realism desired. The more articulate a model needs to be, the more difficult it is to create and (often) manipulate. Solid, immobile objects with well-defined surfaces are the easiest to model, whereas cloth, fur, hair and foliage are more difficult, and motion dynamics complicate things further. The degree of realism desired must also be taken into account. Some films, like *Jurassic Park* (1993) or *Jumanji* (1995), require digital models with a photorealistic degree of realism; others like *The Last Starfighter* (1984) or *The Mask* (1994) are more stylized and less realistic; and films like *Beauty and the Beast* (1991) and *Aladdin* (1992) use digital models whose degree of realism is very highly stylized and unrealistic, although effects of believable perspective and three-dimensionality are important. The degree of realism required depends on the audience's familiarity with the objects being depicted; as Industrial Light and Magic (ILM) programmer Eric Enderton says, "The Holy Grail is to do a believable human in clothes - a human with cloth and hair... This is hard because you know exactly how a human moves, reflects light, and behaves. You've never seen a live dinosaur, which was an advantage for Jurassic [Park]."[31] Computer animation may work best in films where fictional creatures and objects do not have to compete with real-life physical analogs; or, in films like *Toy Story* (1995) where everything, even the human characters, are all computer-generated and equally stylized.

The photorealism usually required of computer animation mixed with live action often requires models to be believably integrated into the live action plates (film images) serving as their backgrounds. Perspective, the matching of light sources for direction, color, specularity, diffuseness, intensity, and a responsiveness to other factors present in the scene, like wind, gravity, and temperature affect the character's appearance or actions. Transparency, translucency, reflection mapping, and environment mapping are other ways of suturing a character into its surroundings by making it appear to interact with the lighting there. With all these things taken into consideration, the cost of animating the model may be equal to or more than that of the physical model. As Greg Jein noted in 1993, "For *Star Gate*, the show I'm on right now, the physical construction costs of the models were $8,000-$10,000 per spaceship, while computer-generating them would cost somewhere in the area of $35,000. If somebody doesn't have that kind of money to play with, they'll find a good modelmaker to do it for them."[32] As the technology develops, costs continue to come down; and in some cases where the digital model could not have been done physically at all, one cannot really say that the digital model is "replacing" a physical one.

While complex things like fire, waving flags, rippling fluid surfaces, and windblown trees rely on the laws of physics (or some cartoon variant of them), character animation requires even more nuance to be believable. Personality, mood, and emotional expression are needed as well as subtle movements that define expression; as cinema is very character-centered, animated creatures must be able to stand up to the most intense audience scrutiny. Even if the illusion of life is achieved, there still is the question of good *acting*, which is difficult enough for human actors (the over-the-top Jar Jar Binks in *Star Wars Episode 1: The Phantom Menace* comes to mind here). Furthermore, as Frank Thomas shows in his essay, "Can Classic Disney Animation Be Duplicated on the Computer?", hand-drawn character animation often diverges wildly from the laws of physics, and the graphic layout itself is designed for composition in two dimensions instead of in three, resulting in the frequent "cheating" in the representation of three-dimensional space.[33] Perhaps the most famous example of "cheating" are Mickey Mouse's ears, which always appear as two black circles, which never foreshorten or overlap, no matter which direction he is facing (nor does one ever get a top view of Mickey).

And despite advances in the digital modeling of natural phenomena, it still seems unlikely that computer animation will be able to fully and completely and believably match and replace the articularity and subtlety human actors are capable of in close-ups, although they can in long shot, if seen briefly (as in *Titanic* and *The Phantom Menace*). Walther Benjamin pointed out how film broke up an actor's performance into shots; and we might add the separation of voice and body, for example, when animated characters are voiced by actors, or when voices are electronically altered (for example, the voices of C-3PO or Darth Vader in *Star Wars*). Digital technology carries the division of a performance's components even further; actor's faces can be texture-mapped onto computer-generated heads (as in *Virtuosity* (1995) or *Lawnmower Man* (1992)), and an actor's movements can control a computer-generated character, through motion-capture technology (as in *Titanic* (1997) or *Jurassic Park* (1993)). Completely computer-generated characters indistinguishable from filmed human beings are still far from being achieved. Equally difficult is the "recreation" of actors from the past; even if resurrecting Bogart and Monroe onscreen were desirable, human impersonators could do a better job. And even if computers *could* do it, the audience would still know they were simulations; although the audience may marvel at the digital legerdemain occurring when Forrest Gump appears to shake John F.

Kennedy's hand, we do not forget that what we are seeing is just a special effect using historical footage of deceased persons.

The subtleties of character animation are not the only ones limiting the use of digital modeling; due to the nature of certain materials, many physical models or effects are very complicated to reproduce through computer animation, or still too expensive and time-consuming to be considered practical. This will undoubtedly change as the technology improves, but audiences will also become more sophisticated in identifying computer effects.

In film production, physical models continue to be used alongside digital models, each filling their niche, doing what they do best. As the cost of digital modeling decreases, it has taken over some of the roles of physical models, but not all of them. The combined use of both methods can contribute to a film by keeping the audience guessing which models were actually physical ones and which were not (when they even notice an effect at all; invisibility is often the goal of the effects team). Special effects films since Méliès have mixed special effects methods from shot to shot the way a magician uses sleight-of-hand to misdirect the audience. *King Kong*(1933) mixed miniatures with full scale mechanical models that interacted with the actors, and even in Méliès' *A Trip to the Moon* (1902) we see full scale vehicles intercut with miniatures. In *Jurassic Park*, the dinosaurs were variously represented by mechanical models, computer-generated models, puppets, and even a man in a dinosaur suit. By varying methods from shot to shot, the dinosaurs appear to be able to do anything, and the audience is kept guessing how it is done.

The processes behind physical and digital modeling are very different, although digital modeling imitates some of the techniques of physical modeling. In order to better represent the changes of form occurring in the movement of the dinosaurs in *Jurassic Park* and especially *The Lost World* (1997), the animators made a computer-generated skeleton, fastened elastic models of "muscles" to it, and covered it with a texture-mapped skin; this way the muscles would be seen deforming the surface of the skin as the animals moved. This method of "building" the model up is similar to those used by the stop-motion animator.

Understandably, some physical modelers are threatened by the rapid technological change, so efforts have been made to convert their work and methods over to digital modeling. As Jeff Mann, director of ILM's Production Operations and former head of the model shop, explained,

> I liked working on a stage with lights, making something look real... There's a camaraderie in the production aspect; you have a common goal to make it real. We worked for ten years to make this process flow

> smoothly, and it seems weird to suddenly do it all on one workstation. The change to workstations is happening so fast.... It's stressful for a fair number of the model builders. ILM is trying to retrain the optical compositors as digital compositors, and to teach some of the model builders to use the tools of the computer to build computer models. Some will be able to adapt, some will get to keep building models, and some will go do something else.[34]

Serendipity often plays a role in the process of making of a physical model. Some of the models for the original *Star Wars*, for example, were built from various plastic model kits mixed together, and were hand-crafted; the same can be said of the puppetry, costumes, and stop-motion models used. The stop-motion feature, *The Nightmare Before Christmas* (1993) had certain nuances, warmth, and feeling that would have been lost had the film been done entirely on a computer. The animators' ability to improvise quickly on the set, and explore camera angles and lighting set-ups, would also be affected.

Certain advantages of physical modeling are incorporated into digital models; some digital models are based on physical ones sculpted by hand as maquettes which are input using laser-scanning digitizers, 3-D electromagnetic probes that record their surface geometry, or through manual digitizing techniques. Photographs are often used for texture, reflection, and environment mapping, and orthogonal photographs displaying the top, front, and side views of a model can be used to reconstruct it in three dimensions. In this sense, the digital model does not replace the physical model so much as it uses it as input.

The process of animating the computer-generated model can also rely on manipulation input directly from the physical world. Performance animation using motion-capture rigs records an animator's movements in real time, transferring them to the onscreen objects. This process is similar to puppeteering, except that instead of a physical puppet, the movements are digitized and fed into the computer as instructions used to control a computer-generated model on-screen. Motion-capture rigs recording movement can be designed to be worn by the animator or performer inputting the movements, or as a table-top model which can be manipulated much like those used for stop-motion filmmaking. The latter design was developed to control the *Jurassic Park* dinosaurs;

> The system [Craig] Hayes developed for *Jurassic Park* (dubbed "DID" for "Digital Input Device") employed a conventional hinge-and-swivel stop-motion animation armature — essentially, a machined metal dinosaur skeleton connected to a Tondreau motion-control system. Although this system was similar to that which Tippett used initially for Go-Motion, the DID differed in that the puppet armature itself was

> never photographed; instead, sensors on the armature fed digital information directly into the computer, where the movements of a wireframe dinosaur on-screen exactly duplicated those of the real-world DID armature.[35]

Although the DID is posed and moved like a stop-motion model, it is still an input device, never seen on-screen, and likely to be reused later. Physical models, however, often have a 'life' outside the film; they are displayed in museums, they appear at conventions, some are bought and sold like works of art or antiques, and some, like the mechanical shark from *Jaws* (1975) or the house from *Psycho* (1960), appear in theme park rides promising a physical, visceral experience, where they are adorned with the 'aura' of the original. This brings us to the extracinematic realm, where digital models are not likely to replace physical ones.

Physical models are more "real" than the digital ones precisely because, like human actors, they have an existence extending off-screen, outside the film. The enormous amount of popular discourse surrounding special effects "magic" thrives on explanations of the intricate design and labor involved in putting effects on film. Articles explaining digital effects methods sometimes show computer-generated wireframe models the way physical effects articles show armatures and miniatures. Still, physical effects and the ingenuity that go into them make up most of the discourse in effects magazines like *Cinefex* and *Cinemafantastique*. *Jurassic Park*'s advertising of its full-size mechanical dinosaurs parallels that of Spielberg's *Jaws* and the press attention given to "Bruce", the mechanical shark that was its star, and there is hardly a description of *Jurassic Park* anywhere that does not mention special effects.

Digital models have already begun to replace physical ones in the cartoon animation industry; since Disney "inks and paints" their features in the computer, there are no actual production cels from any Disney features after 1991 on the market (most character movement in these films, however, originated as pencil drawings on paper; perhaps the sale of these sketches will eventually join cels in animation art stores). Originally physical models were made and photographed to produce images; now the images are made directly, and exist as nothing more than images. Like the shallow facades lining streets of studio backlots, they are hollow shells, surfaces without substance, or perhaps even only images of surfaces. In the case of animated films, characters existing initially only as images are turned into three-dimensional objects for merchandising, completely reversing the process!

As they are integrated together and their niches are redefined, digital and physical effects continue to exist side by side. Digital effects continue to depend on collaboration with physical techniques and input of data from physical sources, since the physical world is where the artist and audience are found. While they may lack certain properties and abilities of physical models, digital models have capabilities beyond that of any physical model; the digital model, after all, is not bound to a three-dimensional existence.

The Leveling of Dimensionality

In the realm of the digital 'object', lack of physicality means a leveling of dimensionality. Everything which is represented in the computer is stored as a series of ons and offs or 'ones' and 'zeroes'; the digital work of art is essentially a one-dimensional entity, whose digits must be read linearly before they can be converted into text, images, computer-generated objects, or whatever they represent. When these 'objects' are stored digitally, they are all reduced to one-dimensional strings of digits; thus digital form levels dimensionality by reducing everything into the same one-dimensional form.

The computer also makes possible conversions between dimensions; for example, in the computer graphics process known as "feature extraction", 2-D images are analyzed visually to produce depth clues, from which algorithms can create a 3-D model; perspective, light and shadow values, reflections, foreshortening and textures enable the recovery of volumetric shape.[36] (Theoretically, a similar process could be done with photographic film; one could take film footage of a tracking shot (the camera tracking sideways at a uniform rate, facing in a direction perpendicular to the direction of motion), and reconstruct a three-dimensional view of the imagery. While tracking, the camera occupies one position for a given frame, and then a nearby position a few frames later; if the speed is right, the distance traveled in the time each frame is taken could be the same as the distance between a person's eyes. If two copies of the film are projected side by side, and their synchronization is offset by exactly the number of frames taken in the time needed for the camera to have moved the distance between a person's right eye and left eye, then the two views would each represent a view from one of the spectator's eyes, resulting in a 3-D film.)

The Massachusetts Institute of Technology's Media Lab has developed "frozen movies", in which two-dimensional images, treated as layers, are stacked, forming a three-dimensional block which can then be sliced in any direction to reveal identifiable patterns.[37] Three-dimensional graphics are used extensively in scientific and mathematical visualizations, for everything from weather simulations, to "phase space" visualizations[38] of strange attractors, to RGB color space, in which any color can be located and numbered according to where it appears along the three axes of hue, saturation, and tone (allowing color to be handled numerically by computer).

Mathematically, coordinate systems are used to describe the sets of points making up lines, curves, planes and three-dimensional solids; every point in space can be defined by a unique combination of coordinates (x, y, z) locating it along three axes, one for each dimension. If a fourth coordinate is added (or more), properties of higher dimensions can be explored —and sometimes visualized— through the use of the computer. One such object that is often visualized is the four-dimensional cube, or hypercube (what is usually being produced, though, is a two-dimensional image of the three-dimensional shadow of the four-dimensional object). Since the first hypercube and other four-dimensional graphics generated by A. Michael Noll in the mid-1960s, computers have been an essential tool in the visualization of four spatial dimensions. Today's computers aid in visualizing even higher dimensions as well as non-Euclidean space, and have opened up the exploration of fractional dimensions lying in between whole-numbered ones, with the objects of fractal geometry. In his book *Fractals: Form, Chance, and Dimension*, Benoit Mandelbrot describes the notion of fractional dimensions;

> . . . physical dimension has subjective basis. It is a matter of approximation and therefore of degree of resolution. . . . To confirm this last hunch, we will take up an object more complex than a single thread, namely a ball of 10 cm diameter made of a thick thread of 1 mm diameter. Depending on one's viewpoint, it possesses (in latent fashion) several distinct physical dimensions.
>
> Indeed, at the resolution possible to an observer placed 10 m away, it appears as a point, that is, as a zero-dimensional figure. At 10 cm it is a ball, that is, a three-dimensional figure. At 10 mm it is a mess of threads, that is, a one-dimensional figure. At 0.1 mm each thread becomes a sort of column and the whole becomes a three-dimensional figure again. At 0.01 mm resolution, each column is dissolved into filiform fibers, and the ball again becomes one-dimensional, and so on, with the dimension jumping repeatedly from one value to another. And below a certain level of analysis, the ball is represented by a finite number of atomlike pinpoints, and it becomes zero-dimensional again.[39]

The dimensional ambiguity of the ball of string is similar to that found in computer imaging; a three-dimensional stereo image is made up of two two-dimensional images, each of which is made up of a zigzagging one-dimensional raster scanline, which in turn is made up of zero-dimensional pointlike pixels.

The dimensional flexibility provided by computer graphics can allow for complex data to be displayed as a kind of "conceptual space", unbounded but finite, its size and resolution determined only by a computer's speed, memory and architecture constructing the "space" being visualized (computer memory is itself already commonly referred to as "space"). In *Cyberspace: First Steps*, Michael Benedikt describes ways of displaying multidimensional data in which objects have *extrinsic* and *intrinsic* dimensions; extrinsic dimensions are the three dimensions of space plus one dimension of time, while intrinsic dimensions are built into the object itself, in its color, size, movement, and position, each of which can be used to display information.[40] Some of these data display objects already have names, for example,

> The choropleth, the use of colour or shading to represent data about geographical regions on maps. The metraglyph, an icon with circular body and attached spikes whose diameter, length and angles can represent data sets. The Chernoff face, a special kind of metraglyph in which the parameters of the geometry of a recognisable object, like a face, represent the data.[41]

Objects in cyberspace can mix Euclidean and non-Euclidean spaces; objects can have unfolding dimensions, and containers can be bigger on the inside than on the outside, and so on. Such "liquid architecture", as Benedikt calls it, is only possible through the digital computer.

The importance of the multidimensional space available through digital technology will always depend on the two junctures it has with the physical world, those points of input and output. Through the process of output, the digital artwork rejoins the analog world as sound, image, or object, either through some display device or as some variety of "hard copy"; printouts, photographs, film frames, or even a three-dimensional object; Cubital's Soldier 5600 can create solid objects, complete with moving parts, from digital models. CAD designs are sliced into layers a few thousandths of an inch thick, and these layers are then joined together, forming the object a slice at a time, allowing three-dimensional prototypes of computer-generated objects to be built automatically. We might say, then, that "multimedia" and "mixed media" are terms which only apply to hard copy and analog media.

The loss of inherent dimensionality is another loss which can be suffered by autographic works when they are digitized (allographic works already are inherently dimensionless). The degree to which inherent dimensionality is a necessary part of the experience of a particular work of art determines the failure of that work of art to be successfully represented in digital form, resulting in a bias against autographic works in general.

The leveling of dimensionality means digitized data of every kind can be manipulated together in the same machine; or that machine translation from one medium to another is possible (once the artwork in question has been digitized, that is). This is the premise behind multimedia, praised as a means of bringing various forms of sound and image together, an incarnation of the *Gesamtkunstwerk* —the fusion of the arts into one all-embracing enterprise, a dream composer Richard Wagner sought after but never achieved.

The digital artwork's lack of physicality enables it to be a nexus point joining many forms of art and media, but at the same time, it is also responsible for the marginalization, exclusion, or adaptation of artwork inexpressible in or not suited to digital form; digital multimedia fulfills the role of a *Gesamtkunstwerk* only if one accepts the digitization that must take place before the fusion can occur. Translation into digital form is not a neutral process (as will be seen in the following chapter), and so, in the end, the project of digital multimedia falls short of being a true *Gesamtkunstwerk.*

NOTES

1. See Bell, T. F., *Jacquard Weaving and Designing*, London and New York: Longmans, Green, and Co., 1895, and Posselt, E. A., *The Jacquard Machine Analyzed and Explained, Third Edition*, Philadelphia: E. A. Posselt, Publisher, and London: Sampson Low, Marston & Co., Limited, 1893.
2. Ibid., page 107.
3. Ord-Hume, Arthur W. J. G., *Pianola: The History of the Self-Playing Piano*, George Allen & Unwin, London, ©1984, page 263.
4. Ibid., page 267.
5. Noll, A. Michael, "The Digital Computer as a Creative Medium", *IEEE Spectrum*, Vol. 4, No. 10, October 1967, page 92.
6. From an advertisement for MicroProse Software Inc.'s *C. P. U. Bach* program, appearing in *Wired*, November, 1993, page 49.
7. Franke, Herbert W., *Computer Graphics - Computer Art, Second, Revised and Enlarged Edition*, Berlin, Heidelberg, New York and Tokyo: Springer-Verlag, ©1971, page 163.
8. See Goodman, Nelson, *Languages of Art*, Indianapolis, Indiana: Hackett Publishers, ©1976.

9. Mitchell, William J., *The Reconfigured Eye: Visual Truth in the Post-Photographic Era*, Cambridge, Massachusetts: The MIT Press, ©1992, pages 50-51.
10. Kubelka, Peter, "Theory of Metrical Film", in Sitney, P. Adams, *The Avant Garde Film Reader*, New York: New York University Press, ©1978, page 159.
11. See Rudolph, Laurence H., "Seeking a Solution to Used CDs", *Billboard*, Volume 105, Number 32, August 7, 1993, page 7; Bach, Russ, "There's No Justifying Used CDs", *Billboard*, Vol. 105, No. 30, July 24, 1993, page 6; and Pohlmann, Ken C., "As Good as New", *Stereo Review*, Vol. 58, No. 12, December, 1993, page 25.
12. Gruber, Michael, "Digital Archaeology: Endangered data, aging archives", *Wired*, November, 1993, page 114. On the vulnerability of digital documents, see Rothenberg, Jeff, "Ensuring the Longevity of Digital Documents", *Scientific American*, January 1995, page 42.
13. Benjamin, Walther, "The Work of Art in the Age of Mechanical Reproduction", reprinted in Hanhardt, John G., ed.,*Video Culture: A Critical Investigation*, Layton, Utah: Peregrine Smith Books, ©1986, page 29.
14. On these restorations, see Talley, Mansfield Kirby, "Michelangelo Rediscovered", *Art News*, 86, Summer 1987, pages 159-170; Beck, James, "New Color on the Sistine Ceiling and Other Issues", *Arts Magazine*, 61, May, 1987, pages 72-73; and Gorlin, Alexander C., "The Sistine Chapel Restored: Cleaning Reveals Unsuspected Colors", *Interior Design*, 55, October 1984, pages 236-241; and Armstrong, George, "Leonardo Revealed", *Art News*, 87, March 1988, pages 162-167.
15. Shulman, Ken, "Like seeing Leonardo for the First Time", *Art News*, 90, November 1991, pages 53-54.
16. Schwartz, Lillian, with Laurens R. Schwartz, *The Computer Artist's Handbook: Concepts, Techniques, and Applications*, New York and London: W. W. Norton and Company, ©1992, pages 278-279.
17. From *Eastman Images, Newsletter of the KODAK Worldwide Student Program*, Fall 1993, Volume 5, Number 3, page 1. The restored film was laser printed onto color intermediate film on a film recorder, and the prints, shown soon after in theaters, were beautiful. (Snow White must have longed for these digital prints when she sang "Some day my prints will come.")
18. Benjamin, Walther, "The Work of Art in the Age of Mechanical Reproduction", in Hanhardt, John G., ed., *Video Culture: A Critical Investigation*, Layton, Utah: Peregrine Smith Books, ©1986, page 33.
19. Ibid., page 30.
20. Ibid., page 29.
21. Ibid., page 33. All of Benjamin's examples here are religious art made as an act of worship; the sculptures invisible to the spectator on the ground are not intended for human spectators, but for God alone. The fact that so much early art was made for religious purposes, its creation often an act of worship, may account for its aura since it represented aspirations higher

than mere material ones. Today, ironically, the work of art has gone from being an object used in worship, to an object of worship itself, for some.
22. Ibid., page 31.
23. Mattick, Paul, Jr., "Mechanical Reproduction in the Age of Art", *Arts Magazine*, Volume 65, Spring 1990, page 64.
24. Each complex number (in the form x + y*i*) is used to represent a point in a plane, and can also be multiplied with other complex numbers. A. K. Dewdney explains the process:

> Begin with the algebraic expression $z^2 + c$, where z is a complex number that is allowed to vary and c is a certain fixed complex number. Set z initially to be equal to the complex number 0. The square of z is then 0 and the result of adding c to z^2 is just c. Now substitute this result for z in the expression $z^2 + c$. The new sum is $c^2 + c$. Again substitute for z. The next sum is $(c^2 + c)^2 + c$. Continue the process, always making the output of the last step the input for the next one. . . The Mandelbrot set is the set of all complex numbers c for which the size [the distance from the origin] of $z^2 + c$ is finite even after an indefinitely large number of iterations.

From Dewdney, A. K., "Computer Recreations: A computer microscope zooms in for a look at the most complex object in mathematics", *Scientific American* 253, August 1985, pages 16-17. See also Bown, William, "Mandelbrot set is as complex as it could be", *New Scientist*, 131, September 28, 1991, page 22; and Dewdney, A. K., "Computer recreations; A tour of the Mandelbrot set aboard the Mandelbus", *Scientific American* 260, February 1989, pages 108-11.
25. Frank, Patrick, "Recasting Benjamin's Aura", *New Art Examiner*, Volume 16, March 1989, page 30.
26. From the *Viewpoint DataLabs Dataset Catalog, SIGGraph '95 Edition*, available from Viewpoint DataLabs, 870 West Center Street, Orem, Utah.
27. The main markets for datasets, as listed in Volume II of their catalog, are: Feature Films, Broadcast Animation, Simulation, Virtual Reality, Courtroom Visualization, Architecture, Multimedia, and Medical Visualization.
28. Savage, George, *Dictionary of Antiques*, New York and Washington: Praeger Publishers, ©1970, page 10.
29. Documents of authenticity packaged with the object, and even the box that it comes in, can be important to the verification of an object's authenticity and are often kept by the serious collector. In addition to serial numbers, Enesco's "Precious Moments" figurines have a symbol or "mark" for identification, whose function is not immediately obvious to the uninitiated;

> The annual symbol, also known as the marking, first appeared on the bottom of pieces in mid-1981. Earlier pieces which had no marking

became known as the now famous "No Marks." The Triangle was 1981's marking. The following years saw an Hourglass, a Fish, a Cross, a Dove, an Olive Branch, a Cedar Tree, a Flower, and finally this year's marking, the Bow & Arrow. The annual symbol tells the collector the year the piece was made. The date on the backstamp is the year the artwork was copyrighted, i.e. you can purchase a piece today that has "1978" on the bottom but in fact was made in 1989 and bears the Bow & Arrow symbol.

(From Silva, Juan Carlos, *Greenbook Guide to the Enesco Precious Moments Collection*, Fourth Edition, East Setauket, New York, ©1989, page 203.) Thus, the potential confusion caused by a serial number, mark, and copyright date, may lead the noncollector into believing a piece has greater value than it does.

30. From the "Electronic Word" column in *Wired*, January 1994, page 35.

31. From Ruckner, Rudy, "Use Your Illusion: Kit-Bashing the Cosmic Matte", *Wired*, September/October 1993, page 76. Also, on the replacement of miniatures, see Christine Bunish, "From Movies to Special Venues: Will Computer Graphics Replace Models and Miniatures?", *ON Production and Post-production*, October 1994, pages 42-47.

32. Magid, Ron, "CGI Spearheads Brave New World of Special Effects", *American Cinematographer*, December 1993, page 28.

33. Thomas, Frank, "Can Classic Disney Animation Be Duplicated on the Computer?", *Computer Pictures*, July/August 1984, p. 20-26. For a more detailed description, see Thomas, Frank, and Johnston, Ollie, *Disney Animation: The Illusion of Life*, New York, Abbeville Press, ©1981.

34. Ruckner, Rudy, "Use Your Illusion: Kit-Bashing the Cosmic Matte", *Wired*, September/October 1993, page 76.

35. Magid, Ron, "After Jurassic Park, Traditional Techniques May Become Fossils", *American Cinematographer*, December 1993, page 60. Also, on motion capture, see Christine Bunish, "Snatching the Subtleties of Movement: Motion Capture Gets Even More Sophisticated", *ON Production and Post-production*, November 1994, pages 42-47.

36. See "Efficient Recovery of Shape from Texture" by Davis, Janos and Dunn in the anthology *Digital Image Processing and Analysis: Volume 2: Digital Image Analysis*, Rama Chellappa and Alexander Sawchuk, eds., IEEE Computer Society Press, © 1985.

37. Kevin Kelly, "Frozen movies", *Wired*, Premier Issue, page 17.

38. A "phase space" is a graphic visualization relating possible states of a system in a two or more dimensional representation. Certain variables of the system are chosen and used as the dimensional axes; points appearing in the phase space, then, each have a unique set of coordinates which represents a specific state of the system when the variables used are set at the represented values. Phase space visualizations are useful for conceptualizing how various states of a system are connected, cycles that it might move through, and boundaries of the system's behavior.

39. Mandelbrot, Benoit, *Fractals: Form, Chance, and Dimension*, San Francisco: W. H. Freeman and Company, ©1977, pages 19-20.
40. One example of a multi-dimensional object he gives is;

> A tumbling arrow of variable length is (visually) an eight-dimensional object: 4 extrinsic (3-space plus time), and four intrinsic (3 angles for the direction of the arrow, and 1 for its length). A quivering, tumbling cube of changing color is a 16-dimensional object: 4 extrinsic again, and 12 intrinsic (3 for angular orientation, 6 (at least) for the amplitude and frequency of the quivering face pairs, and three for the color (say, RGB values)). Clearly, it is possible to "see" a surprisingly large number of dimensions at play before the percept becomes unfamiliar.

Benedikt, Michael, "Cyberspace: Some Proposals", in *Cyberspace: First Steps*, edited by Michael Benedikt, Cambridge, Massachusetts, and London, England: The MIT Press, ©1992, page 142.
41. Mallen, George L., "The Visualisation of Structural Complexity: Some Thoughts on the 21st Anniversary of the Displays Group", in *Computers in Art, Design and Animation*, John Lansdown, and Rae A. Earnshaw, editors, New York: Springer-Verlag, ©1989, page 24.

4.
Cultural Biases Inherent in Digitization

Digitization, as a form of translation, is not a neutral process, for it changes whatever passes through it. An object's physical nature is lost in translation, and even sounds and images from other media are not immune to loss and changes in form and content. Although the flexibility of digital form makes algorithmic translation between media possible, certain biases within the technology make certain types of works better suited to digital media than others. In some cases, these biases are easily remedied by designing technology with a variety of cultural needs in mind (particularly in the area of word processing). Other biases, however, are not so easily amended. For example, it cannot always be assumed that there is enough resolution and memory for an adequate digitization; budgeting limitations (regarding either money or memory) often limit what is available. The means of digitization can also vary, especially for the digitizing of three-dimensional objects. And in all forms of digitization, the question of "How much resolution is enough?" is often the subject of debate.[1]

Even given enough resolution and memory, many things cannot be adequately digitized or represented in digital form. To begin with, digital media cannot yet reproduce smell, taste, or certain aspects of touch (some force-feedback datagloves can crudely represent rougher textures, but not much more —and the material of the glove itself will limit what it can emulate). Visual and sonic art fare better, but still fall short of what the human sensorium is capable of sensing. As in analog media, the context of the work of art is also often lost; the experience of standing and looking at a *trompe l'oeil* painted on a wall to match the perspective of the surrounding architecture is a much different experience than seeing a reproduction and reading about its effects and positioning. The same is true for voice and piano recitals, where performers often adjust their sound according to the acoustic space of the performance. A digital artwork is more easily detached from its context than an analog one, since it does not rely on physical variables, it is less firmly bound to its point of origination, and there is often no indication of age or history apart from the file format and content.

The shift from analog media to digital media on a large scale is relatively recent and its biases are often left unacknowledged or unexplored. Some media-specific biases have already been hinted at in past chapters, such as the favoring of allographic works over autographic works, and the favoring of two-stage works over one-stage works. Other biases stem from the fact that cultures differ widely in regard to the importance of various media; for example, while the written word was of central importance in the cities of Western Europe, the empire of the Incas, with a population of approximately ten million, was organized without any form of written language. Media technology is a product of the culture that creates it, a point raised by Junichiro Tanizaki in his 1933 book, *In Praise of Shadows*. Discussing Japanese culture, he writes;

> We would have gone ahead very slowly, and yet it is not impossible that we would one day have discovered our own substitute for the trolley, the radio, the airplane of today. They would have been no borrowed gadgets, they would have been the tools of our own culture, suited to us.
>
> One need only compare American, French, and German films to see how greatly nuances of shading and coloration can appear in motion pictures. In the photographic image itself, to say nothing of the acting and the script, there somehow emerges differences in national character. If this is true even when identical equipment, chemicals, and film are used, how much better our own photographic technology might have suited our complexion, our facial features, our climate, our land. And had we invented the phonograph and the radio, how much more faithfully they would reproduce the special character of our voices and our music. Japanese music is above all a music of reticence, of atmosphere. When recorded, or amplified by a loudspeaker, the greater part of its charm is lost. In conversation, too, we prefer the soft voice, the understatement. Most important of all are the pauses. Yet the phonograph and radio render these moments of silence utterly lifeless. And so we distort the arts themselves to curry favor with the machines. These machines are the inventions of Westerners, and are, as we might expect, well suited to the Western arts. But precisely on this account they put our own arts at a great disadvantage.[2]

Even when media originate within a culture, the order of their appearance influences the form they take. Due to the timing of their development, early film was sometimes used as a vaudeville act, sometimes imitating its content, and many early television shows were visual extensions of radio programs. Digital media are still imitating and integrating earlier media like video and film, alongside newer forms like searchable databases and CD-ROMs. As the centrality or marginality of any particular medium varies from culture to culture, these media-related biases may become cultural biases; for instance, an orally-based culture will be more readily adaptable to the sound and images of digital media than one based on tactile means of expression or olfactory symbolism.[3] Economic, graphic, textual, and oral biases whose effects are often culturally specific are also present. The first section of this chapter asks what economic background is needed to support digital media, and the second examines biases within the nature of the media themselves. The last section is a brief look at the digital culture arising from applications of the technology as a culture all its own, and some of the ways it differs from traditional cultures.

The Economics of Representation

The importation of digital media into a culture means the installation of entire technological systems, if they are not already in place. Unlike the printing press, signboards, or musical instruments which could be of use by themselves, digital media usually require a steady supply of electricity, which will usually come from government or industry (even if batteries or solar cells are used, they still have to be purchased from somewhere). In certain climates such as the tropical and sub-tropical, electricity may also be needed to maintain conditions for optimal hardware operation, such as air conditioning and humidity control for computer rooms. Whether the electricity comes over lines from a central power source, in the form of batteries, or even generators, there will be a market for big business or government to fill, and users of the electricity will have lost some of their independence in the process.

The dependence on technological systems does not end with electricity; a system of technological experts must be kept on hand for the installation of the equipment and its maintenance and repair, for the constant training of the equipment's users, and retraining as newer software and hardware upgrades emerge. Some types of architecture are better at accommodating the installation of computer systems than others; as David Lammers points out, few traditional Japanese buildings have the wire closets, dropped ceilings, or floor ducts required for installing a fiber-based network.[4] A steady stream of material supplies will also be needed; printer paper, toner, diskettes, and so on, as well as new parts and machines when the old ones break down. In some cases, entire industries will rise up to fill these needs, greatly changing a country's economic infrastructure, as can be seen in a number of Pacific Rim countries in Asia. Some industries, begun initially as a supply of cheap labor to Western countries, preceded the influx of digital media into the Far East and can even account somewhat for its presence there. In this sense, it is perhaps inevitable yet ironic that Asian firms are competing with and surpassing American firms in some electronics markets.

Much digital media equipment is high-end and still relatively expensive, especially when import tariffs are added. Authoring software, needed for media production, is another expense, requiring additional memory and hardware such as image scanners, video cards, laser printers, and so on. Digital media are still much cheaper to consume than to make and distribute. Even in the area of home exhibition, great disparities can exist, due to the quality of the players or equipment being used. Whereas people of varying economic backgrounds attending a live concert performance will enjoy roughly equal sound quality (differing, perhaps, only in the distance and location of their seats), the sound quality of a recorded performance on a home system will depend largely on the equipment used (and the amount of money invested in it). Indeed, "performance" is now often used to refer to the operation of audio equipment itself as well as the human performers.

Although education is often a reason for the installation of computers, a certain amount of education is already required just to use them; written language ability, and quantized and procedural styles of thinking are often necessary for successful computer usage.[5] Once systems have been converted over to the technology, users are forced to learn it; many public and university libraries have holdings information available online, and books are no longer filed in card catalogs; only those familiar with the information needed to use information technology can get at the information available through it.

Since digital technology is so pervasive (appearing in everything including watches, washing machines, automobiles, toys, and cooking appliances), its application can encourage a culture's growth in machine dependence, although the value placed on technology varies greatly from culture to culture. The Amish are often seen as one of the cultures most resistant to technology, aware of its potential harm to social order, and careful to separate out its positive and negative aspects. Paul Levinson writes;

> ...no Amish group has rejected technology outright - rather, they struggle with the appeal of technologies, usually accepting a new machine at first, agonizing over its real and projected social consequences, and then deciding whether and to what extent it will be used. They come the closest I have seen to a living embodiment of a philosophy of technology.
>
> In most cases, the Amish in fact accept new technology, while straining the limits of creativity to keep it from disrupting their social order. Electricity from central power companies is forbidden - sockets in the wall are appendages of an uncontrollable, huge, external, political-economic power structure - but electricity from 12 volt, self-

> sufficient batteries is all right. ... The Amish proscription on centrally supplied electricity dovetails nicely with its religious-ethical dislike of mass media - like television, which usually can't run on batteries - and its content. But a new device has crept into some Amish homes and businesses: a clever little "inverter" which transforms 12-volt battery current into a reasonable likeness of the 110-volt power that comes from the socket. ... Pocket calculators that run on batteries and sun are already in widespread use. As far as I could see, laptop and notebook computers are not yet used, but the profound decentralizing effect of these media makes them natural allies of the Amish. ... How intriguing to think that the Amish, and their deep sense of privacy and independence, could be a source or even a spearhead for enlightened digital technology to empower the individual and the pioneering community in the next century.[6]

Although it may be too soon to gauge the effects of digital technology on the Amish community, Levinson does not note how the dependence on manufactured goods (the calculators, the "inverter", and the batteries themselves) diminishes the independence of a community that traditionally did all of its own farming and building, and even produced its own clothing. For the average American reliant on technology these media may have a "decentralizing effect", but for the already decentralized Amish, they may become a centralizing force and a compromise of independence.

Whenever one culture becomes dependent on another, particularly on one bent on furthering its own ends, there arises the opportunity for cultural imperialism. The sale and installation of digital media and the technological systems they need are an obvious example, as the rise of the computer industry coincided with the opening up of foreign markets to American advertisers, during the early 1960s. But the means of cultural imperialism need not always be so overt. The hiring of a foreign technical expert, for example, can often involve the importation of a hidden cultural agenda. As George Foster puts it;

> The combination of fine technical training and an ethnocentric point of view leads to false and dangerous definitions of a good technical aid program and the role of the international technical specialist. The "good" program or the best technical assistance comes to be defined as the *duplication of American-style programs and projects*. The obvious corollary is that the best technical expert is the person who *most perfectly transplants an American-style program.*[7]

Cultural traditions can also be restricted by technological limitations: in Japanese culture, one cannot bow or avert one's eyes online, and in England, the importance of accent is lost on the net. Also, in Japan, some men do not like to use keyboards because many Japanese men still consider typing to be a woman's job, an attitude which slowed down the adoption of microcomputers.[8]

In this age of electronic media, it is no longer even necessary to send human agents to a foreign country in order to infect it with American culture; electronic signals sent by satellite do just as well, coupled with the capitalist urge of Western-inspired locals who gain wealth and position by assisting the foreign invasion, in what amounts to a form of electronic neocolonialism. Jeff Greenwald describes the situation in India;

> Along with a few similar electronics bazaars across the Indian subcontinent, Lajpat-Rai is a crucial link in this vast republic's exploding market for satellite television - and a central supply port for a new generation of opportunistic entrepreneurs called "dish-wallahs" - *wallah* being a common Hindi phrase which translates to something between "hack" and "specialist". Using no-frills satellite dishes, simple modulators, and hundreds of meters of cable, these inventive television hackers are affecting a transformation that many locals feared would never come: propelling Mother India out of information limbo by hard-wiring its living rooms directly into the global jet stream of satellite news, live sports, and the geosynchronous gyrations of MTV. ...With 60 apartments hand-wired to his rooftop dish, [Deepak] Vishnui may be one of the smallest dish-wallahs in Delhi, but his operation is completely typical. With subscriber bases ranging from 50 to 1,500, all of Delhi's 200 to 300 cable networks (a fraction of an estimated 20,000 scattered across India) are privately owned, bare-bones businesses with tiny overheads. . . . "There is no doubt," said a salesman for a New Delhi dish-wallah, "that anyone with 25,000 rupees (less than $1000) can walk into the Lajpat-Rai market and walk out with everything they need to build their own cable television station."[9]

Despite having a strong film industry (the largest in the world) that has avoided domination by American films, the free TV signal is eagerly accepted by those who watch it in their homes as well as the entrepreneurs who profit from it. The effects, however, are beginning to be seen;

> "There is a problem," admits Vinod Tailang, president of the 120-member Cable Television and Dish Antenna Operators Association of Uttar Pradesh, and father of two teenagers. "Much of what we show is totally against Indian culture. It is definitely changing the way Indian boys and girls react to each other. I see them dancing and moving all the time. Jumping, jumping. They have become action-packed."[10]

Interestingly enough, Tailang says "what *we* show", acknowledging his complicity in the cultural shift at the same time that he critiques it.

Complicity comes not only through the importation of cultural content, but through the means of delivery of that content, the media forms themselves. Not only do other cultures see and learn about mainstream American culture through digital technology, in some cases they learn about their own culture through it. Although technically speaking, the Inupiat Eskimos of northern Alaska live on "American" soil, their cultural is undeniably distinct and much older than what we might term "mainstream" American culture. As Jerry Franklin reports;

> Scattered across the barren 88,000 square miles of Alaska's North Slope Borough - the largest and northernmost county in the US - are eight tiny towns. Most of the residents are Inupiat Eskimo. Most still hunt and whale, and many speak little or no English. Here there is no infrastructure... Here, also, thanks to an ambitious school district and millions in oil-derived revenue, is a pioneering digital classroom for many of the Borough's 2,000 public school students. ...Two studio cameras bring the instructor, text, and graphics into each remote site. Along the way, the signal is digitized, compressed, multiplexed, microwaved, and up- and downlinked via full-time dedicated circuits. Instructors have two monitors - one to see themselves, the other to see the classrooms, which are shown one at a time, security-monitor style. When a student in one of the classrooms speaks, the camera pans to the source of the voice, so everyone can see who is talking. Distance Delivery allows the Borough to bring courses like trigonometry to the far-flung villages. This term's most popular class, however, is Inupiat Studies. The irony of using digital-age technology to teach age-old tradition was brought home one morning when a non-English-speaking village elder was brought in as a guest lecturer; much class time was spent teaching the elder how to use the remote control to address specific sites.[11]

That the natives of a traditionally non-technological culture should have to learn about their own culture in such a disembodied and heavily-mediated way is a sad statement of how dependent the transmission of cultural tradition has come to depend on digital media (and, indirectly, on the culture producing the media). And, as in the case of the elder, the technology will often have to be learned *before* it can be used to transmit cultural knowledge; mastery of foreign cultural artifacts will precede the transmission of native culture. Nor does culture recorded, preserved, and transmitted through digital media remain unchanged, for within the technology itself there are cultural biases built into the means of representation.

Economical Representation

Even if digital media *were* affordable and accessible to all, cultural biases would still exist due the nature and form of digital representation. Of course, all forms of media contain biases to some degree; even print media could be said to be biased since it is of use only to cultures with a written language. As we have seen in chapter one, the bases of digital technology, quantization and digitization, are themselves products of a certain cultural world-view, so it is not surprising that digital media should also reflect their cultures of origin.

There are two main components of digital technology from which restrictions and biases emerge; the grid and the memory. The grid is where the digital elements are positioned; graphically, it appears as the plane of pixels making up the image, and sonically it appears as the series of quantized levels and samples approximating a sound wave. The sizes of these grids depends on the scanning or sampling rates used, which in turn is limited largely by the amount of memory available. Computer memory, wherein a finite amount of data can be stored, can limit resolution, length, and complexity, since an increase in any one of them will usually mean that more memory is required. Computer memory, as the currency of digital media, is always in demand; and financially speaking, more memory (or more bandwidth, in the case of transmission) costs more money. Thus, to keep costs down while making the most of the available memory, strategies such as lower resolutions and compression algorithms are designed to make digital representation as economical as possible.

The desire for economical representation favors representations using lower resolution taking up less memory, rather than higher ones providing greater detail. In graphic art, this results in a bias towards horizontal and vertical graphic elements over curved or diagonal ones, since the grid arrangement of the pixels becomes more noticeable as resolution decreases; lower resolutions emphasize the horizontal elements of curved and diagonals, and their quantization on the grid results in jagged series of stair steps, known as spatial *aliasing*. Other effects like the "ringing" in the video image are also due to the underlying grid. "Ringing" occurs when a pattern depicted onscreen and the grid of the pixels making up the image interact and create a moiré pattern (clothing with very thin stripes or venetian blinds in a scene often cause ringing).

In an image, limiting the number of hues and tones used (the color resolution) will also lessen the amount of memory needed. The effects of limited spatial resolution and limited color resolution can contribute to a bias against the use of depth perspective and towards a flatter, more perspectiveless image, since varying levels of detail (and changes in color saturation) are often used as indicators of depth in graphic art.[12] By reducing effects of depth and perspective, the realism of the picture, and its effect on an audience, is reduced as well.

Thus the designs found in weavings, quilts, and certain Mondrian compositions translate well into digital form, requiring less memory and resolution than, say, the works of Rembrandt van Rijn or Henry Moore. Whole movements may adapt better to low resolutions than others; the graphics of abstractions like Joseph Albers' *Homage to the Square* or Malevich's early Suprematism (which used only squares) would translate fairly well into digital form, since minimalism generally needs less resolution, whereas baroque or rococo works would generally require more. (Of course, some would argue that minimalist works rely on extremely slight variations in tone and color which are barely perceptible and are made more perceptible by being the only variations present; if this argument is taken, than the argument can be made for digital media being biased against them; the point is that some works translate better than others into digital media.) This is not to say that certain works or styles should be translated into digital form while others should not; rather, different works will require different resolutions to be reasonably represented, and thus some will be more "expensive" than others. On the other hand, it also means that some works are more likely to appear degraded because of insufficient spatial, tonal, or color resolution, since images of different types will often be required to fit into the same standardized formats, screen displays, printer resolutions, etc. The images may even be scaled down further for economy's sake, losing most of their subtleties in the process; this is similar to museum catalogs which often depict all the works of a show as black and white photographs. While black and white works (like woodcuts) reproduce well, paintings with subtle uses of color will lose more in translation.

Image compression schemes are another way to attempt to save memory, although they are also often "lossy" processes in which image quality is exchanged for memory saved. (There are some exceptional compression schemes, though; Michael Barnsley's fractal compression schemes work amazingly well in maintaining image quality, allowing for extreme compression with relatively little loss.) Compression algorithms generally work by locating areas of an image that can be stored more economically in memory; areas of solid colors and repeated patterns, for example, can be represented by a smaller sample of pixels than areas of greater detail and variation. Some forms of compression represent an additional quantization of the image; JPEG compression uses Discrete Cosine Transform (DCT) a special case of Fourier transform, converting the pixel data into a set of frequencies and amplitudes, which are then quantized to save memory, and Recursive Vector Quantization (RVQ) compression breaks the image up into tiles, which are then quantized, meaning that only a limited number of different tiles can be used to reconstruct the original picture. Different compression techniques have different effects on images, and there are even programs like Doceo's *Video Compression Sampler* which allow a user to compare their effects.[13]

In analog media, two photographs of the same size and type will cost the same to store or reproduce, regardless of content, but as compressed digital files they will vary in cost, because one might require more memory to be stored than another, due to differences in how well they compress (or, they will take up the same amount of memory but one may be more degraded than the other). Similarly, the repeated patterns found in weaving, tilework like that of the Alhambra overlooking Granada, Spain, or the tessellated works of M. C. Escher would likely compress better than the intricate chaos of a Jackson Pollack action painting or the shimmering colors of a Monet. The repetitions found in M. C. Escher or Islamic art are even seen by some as precursors of computer art; as Mike King writes,

> Islamic art is a good example of visual expression that relates very closely to certain types of geometry—the tessellations of the Euclidean plane. Since the time of Pythagoras, on whose geometries Islamic geometry is largely based, mathematics has developed tremendously, and with computer graphics the visual expression of these ideas has become commonplace. I see the development of computer art as being closely related to the use of these geometries, both old and new.[14]

Escher's tilings of the plane, in fact, have inspired Intellimation's Mac-based *Escher-Sketch*, a computer program that allows users to create their own plane-filling shapes. But the greatest repetition in artwork, as far as individual images are concerned, can be found in moving imagery, where consecutive frames are often very similar. In order to reduce the large amount memory needed to store moving imagery, video compression schemes are used.

Compression algorithms are often needed to import video sequences into multimedia computer programs; Apple's *QuickTime*, a video clip player, is included with most Macintosh computers that can run it. *QuickTime* allows the user to compress video clips at a variety of compression rates, depending on how much memory is available. Video compression, in general, works in a similar manner as other forms of image compression, and since there is often little change from one frame of film or video to the next, a sequence of images is almost like a repeated pattern. Video compression programs, known as "codecs" (short for coder/decoder), can use either *intraframe* or *interframe* compression, or both. *Intraframe* compression mathematically simplifies areas within a frame of video, reducing the memory needed to store the image. *Interframe* compression saves only the differences (the "delta") between frames; a few key frames are stored, and then, instead of storing all of the frames that come between them, only the differences of those frames from the key frames are stored. Thus, sequences with lots of movement or with camera movement will not work as well with interframe compression as well as scenes with less motion, like a talking head, will. Still, no coding scheme is perfect;

> There is a potential downside of any video compression scheme: All codecs use a combination of lossless and lossy techniques to squeeze the video, meaning that the video that comes out isn't an absolutely exact copy of the video that goes in. Some video information is discarded, some is replaced by mathematical approximations of the original. ... In general, the greater the amount of video compression, the greater the amount of loss.
>
> Of course, most codecs give you a great deal of control over the amount of loss. With most systems, it becomes a tradeoff. If you want a large image, you may have to sacrifice color depth (e.g., use 256 or 65,000 colors instead of 16.7 million) or use a slower frame rate. If image quality and full color are paramount, you'll have to live with a smaller on-screen image.[15]

The "smaller on-screen image" means low resolution, which will reduce image quality, as well as other things (see chapter ten, on indexicality, for more detailed discussion pertaining to changes occurring in the pictorial, semiotic, commercial, and psychological values of an image when it is translated into digital form).

Obviously, there are biases towards certain styles inherent in this technology. Cinematically, the static shots of Yazujiro Ozu would compress better than the moving camera of Miklós Jancsó; the uncluttered frame of Antonioni would compress better than the cluttered frame of Terry Gilliam; and the Bazinian long take would generally compress better than the Eisensteinian montage sequence. Andy Warhol's *Empire* (1964) would compress extremely well.

Again, differences in compression are not keeping any films from being digitized, but some films will have greater trade-offs to make when computer memory runs low (no matter what the state of future media, it is a safe bet that *Lawrence of Arabia* (1962) will always lose more in translation than *Arnulf Rainer* (1960) will). Graphically, these biases amount to a preference for a simplified and abstracted image; simple, repetitive, static, and drained of subtlety. The tiny, posterized, stuttering images of *QuickTime* clips can only hint at the works they represent; and they potentially distance the viewer from the works themselves as well as what they represent. These problems, which increase as resolution decreases, are still rather limited in scope since the scanning and digitizing hardware and the large amounts of memory needed mean that most high-resolution image processing takes place in institutions like universities and businesses, with few home users able to afford it (much less the poor of undeveloped countries). And the problems and cultural biases in video and film imaging are small compared to those in what is perhaps the most common kind of computer graphics throughout the world; text characters.

Text Displays

Some of the greatest cultural biases of digital media are found in the area of text display. Resolution is again a problem, along with input, output, software, and character sets. Word processing by computer was, from its very beginning, largely an Anglocentric and Eurocentric venture. In the May 1990 issue of *Communications of the ACM*, editor E. H. Sibley wrote;

> In keeping with our new editorial focus for Computing Practices, we are presenting our first 1990 special section, "Alphabets and Languages". In thinking of a suitable editorial introduction, I realize that I have been surprisingly naive about the problems faced by a large portion of the world's population using Latin font typewriters— and thus of the need for computing terminals, other interface devices, and software for special scripts or languages. I knew, of course, that the Japanese had been developing special terminals for their industry and that the Chinese spelling of some names and places were changed to aid in typesetting, but I still felt that the QWERTY keyboard and ASCII font was enough. ... It became obvious to me that computing in another language would not be as easy as it might at first seem. It is not just a word-for-word translation or transliteration—not even for simple concepts such as column headings. In the simplest of examples, the words may not need to be written left-to-right and/or down, nor may they need the same space to convey the same message. Even a dot matrix printer may prove difficult to use, if not totally inadequate![16]

Indeed, as one moves away culturally and geographically from the English-speaking nations, through Europe to the Middle East and the Far East, Anglocentric biases become evident. The computer, as a word processor, was designed for English, as was the ASCII (American Standard Code for Information Interchange) character set; other countries using the Roman alphabet like France, Germany, Norway, Spain, or Poland require diacritical marks and even additional characters not found in English ("Ç", "ß", "Ø", and several others which are *not* represented in the extended character set of the Macintosh computer that I am using to write this). What is even more telling is that while a number of European characters are omitted, other characters, like "®", "©", and "™" *are* included, indicating the importance placed on the commodity in American culture. And, of course, Apple's own character is included.

Moving further away from English, there are the Cyrillic, Greek, and Hebrew alphabets, and finally Arabic and Asian languages some of which are not even alphabetical. Nor are many of these languages read in the same left-to-right and top-to-bottom order as most Western languages. While programs exist for a variety of languages, digital word processing is still better suited to European languages than those of the Middle East and Far East, since word formation consists of discrete units (letters) printed sequentially. Arabic, the mother tongue of some 200 million people in 21 countries, is more complex graphically, requiring more than just an extra character set;

> A large number of diacritical signs, which are similar to accent marks in European languages, are used to mark short vowels and emphasize or loosen a letter's sound. These marks can be mixed and written above or below the characters to produce composite phonetic effects.
>
> Arabic script, which evolved in contemplation of the traditions of handwriting, is context sensitive. The shape of most characters depends on their position within a word and the characters adjacent to them. Each character may be represented up to four different ways of which only one would be correct in a particular situation. Moreover, many of these intermediary forms depend on the adopted calligraphic style. As an extreme case of contextual reshaping, Arabic script allows ligatures between characters. In other words, adjacent letters can be fused to produce new graphical forms. In order to emulate handwriting, approximately 400 ligatures are available, yet only three of them have mandatory usage.
>
> Orientation of the Arabic script is from right-to-left, but Arabic numerals are written and read from left to right. Unlike Latin-based alphabets, elements of the Arabic script convey directional semantics, which control the orientation of a typesetting process...[17]

Chinese characters are traditionally read in columns rather than rows, although modern business usage has adapted to the left-to-right orientation of Western languages to adapt to Western computer systems. Whereas dot matrix printers were designed to print row by row, the recent change over to laser printers will allow for vertical orientation to return, at least in the printed form.

Laser printing is also an improvement because its resolution is much higher than dot matrix printers. Arabic, as well as Far Eastern languages like Chinese, Korean, Thai, and Japanese, require greater resolution because of their graphical complexity. Whereas a typical IBM-compatible system displays roman alphabet characters in a 9 x 16 pixel cell (which includes the space around the character as well as underneath), Chinese, which has the most complex characters, is often standardized with characters confined to a 24 x 24 pixel cell, and dot matrix printers, previously 9-pin, are now made with 24-pin heads.[18]

The higher resolution takes up a lot of memory, and so does the character set, due to the enormous number of characters that occur in written Chinese. According to one estimate, "In creating a character font, space should be reserved for about 8000 characters. ... For 24 x 24 resolution, about 576 Kbytes is needed."[19] The enormous size and nature of the character set also make inputting difficult. Unlike Western languages or even Arabic, Chinese characters are not composed of letters arranged sequentially. Japanese complicates the matter further because it uses several different kinds of writing, *including* Chinese;

> The first writing style is kanji (writing of the Kan dynasty of China). This is the traditional and formal style of writing. The second kind of writing is katakana, a phonetic alphabet used to spell out words that are foreign to Japanese. The third kind of writing is hiragana, an informal, abbreviated style. Japanese computers have to display a mixture of kanji, katakana, hiragana and ASCII roman.
>
> Japanese characters are usually entered phonetically, often spelled out using the anglicized version of the character. For example, to enter the character for *sushi*, the user enters S-U-S-H-I. Simple for us, but not intuitive for Japanese. Another degree of sophistication hides behind all of this: the pronunciation of a word or character depends on where and how it is used in a statement. As with Arabic, characters change on either side of the cursor as the user enters text. If the word isn't resolved to kanji or hiragana, it will end up either katakana or roman.[20]

Originally, typewriter keyboards for the inputting of Chinese characters had thousands of keys; a two-dimensional selection array for a typical Japanese typewriter was an enormous grid, made up of 35 rows of 63 keys each. Because working with such huge arrays made touch-typing a slow and difficult process, other options were explored for computer input, including writing of characters on a graphics pad with a light pen, voice recognition, and methods which break characters up into sections or strokes which are then built, by multiple keystrokes, from component parts.[21] These explorations resulted in dozens of input systems, each with its own problems keeping it from being chosen as nationwide standard; in 1985, Wei Cui wrote that, "There are now two or three hundred different methods for designing a Chinese character keyboard, but no single method has gained unanimous acceptance, nor have standard design criteria been established."[22]

While Chinese characters (known as *hanzi*) cannot be easily divided into linear elements, the syllables of the spoken word can, since all speech is necessarily linear in form. According to Joseph D. Becker;

> The Chinese language poses problems for phonetic conversion typing because it lacks a traditional phonetic script and a consistent national pronunciation. A phonetic typing system must provide variant dictionaries for each of the major Chinese "dialects" (Mandarin, Cantonese, etc.), although Mandarin, with its 750 million native speakers, is a good place to start. As we will see, standard Chinese phonetic spelling systems do exist that are internationally recognized and easier to learn than any set of hanzi coding rules. Unfortunately, even native speakers of Mandarin often depart from the approved "standard" phonetics. The unavoidable fact is that persons who wish to type Chinese must first take the time to learn something, and in Chinese society their time is far better invested in learning standard Mandarin pronunciation than in memorizing some computer company's ad hoc set of coding-scheme rules.
>
> In the final analysis, the phonetic conversion method is hardly "ideal" for typing Chinese—it is merely superior to any of the alternatives. Indeed, because hanzi typing is so complex, even phonetic conversion systems require careful software engineering to be usable at all.[23]

Because the number of different syllables used in Chinese is still too large for a one-to-one correspondence with a keyboard, the syllables are broken down further into initial and final elements.[24] These phonetic elements are assigned to keys on a keyboard and labeled with their representations in the roman alphabet; this system is known as the "pinyin" keyboard.

The choice of Mandarin pronunciation over arbitrary computer-coding schemes is culturally sound, yet not entirely free of foreign influence either, since familiarity with the pronunciation of the roman alphabet is needed in order to read the pinyin keyboard in the first place. To type in Chinese, a user must first phonetically translate each word into a foreign system, before the word can be entered. This would be equivalent to English-based computer keyboards having keys labeled in kanji which could produce English words if combined correctly.

Suggestions have been made that attempt to lessen the language bias (for example in "Making Languages English Independent", A. Skjellum describes a set of symbols which could replace C compiler keywords, making programming less Anglocentric),[25] but the adaptation of foreigners to digital technology often seems to happen faster than adaptation of technology to foreign markets, although the latter has been catching up in recent years. Part of the reason for this willingness to adapt (on both sides) are the great advantages —and commercial market potential— that come with the computer. Since the computer makes word processing more accessible than the typewriter, digital technology's possibilities are unavailable to analog media. As Joseph D. Becker pointed out, "Ironically, the Koreans, who launched typing technology with their invention of movable type, have had no adequate way to type their language until the recent advent of computerized "removable" type."[26]

There is also opposition to the cultural compromises that can occur; the issue of simplified Chinese characters is a controversial one in Taiwan, where the government created a prohibition against simplified characters to preserve traditional Chinese culture.[27] Nor do all biases favor the West; digital technologies developed in other countries can be just as culturally specific. In regard to the Japanese development of the fax machine, Nicholas Negroponte of the MIT Media Lab writes;

> As recently as ten years ago, Japanese business was not conducted via letter but by voice, usually face to face. Few businessmen had secretaries, and documents were written, often painstakingly, by hand. ... The pictographic nature of Kanji made the fax a natural. Since little Japanese was then (and is now) in computer-readable form, there was (and is) no comparable loss. In a very real sense, fax standardization, lead by Japanese companies, gave great short-term favor to their written language, but resulted in great long-term harm to ours.[28]

The "harm" to English that Negroponte is referring to is the fact that fax transmissions are not directly computer-readable, but rely on character recognition instead. In light of the Anglocentric nature of computing technology, such a complaint seems like the pot calling the kettle black!

The real irony in the whole debate of the international compatibility of text displays is that while Chinese syllables are broken down into sounds represented by the roman alphabet on the pinyin keyboard, English-based computing has come to rely extensively on the use of icons, which in many ways are similar to Chinese pictographs. Although this does not necessarily point to a cultural convergence, it does indicate how technology can be used to level differences and attempt to build bridges between cultures, while at the same time carrying the potential for cultural imperialism, even in an inadvertent manner.

Other means of unifying international information exchange include proposals for a character set combining characters from different countries into one universal set with a standardized code for digital representation. This would allow more people in different countries to communicate by electronic mail, and make it easier for software publishers to develop applications using different languages.

There have been attempts at an international character set. In 1991, twelve hardware and software companies, including Apple, IBM, Microsoft, Sun Microsystems, and Xerox, formed a consortium and developed a universal character code they called Unicode. Unicode was designed to replace the ASCII code, and it also supported Greek, Cyrillic, Armenian, Hebrew, Arabic, Ethiopian, Sanskrit, Bengali, Thai, and other alphabets, and included *some* Chinese, Japanese, and Korean characters (though not all of them).

Soon after its organization, the Unicode consortium was involved in a dispute with the International Standards Organization, whose own multibyte code (known as ISO 10646) had been in development for six years; their code, at 32 bits per character, was twice the size of Unicode's 16 bits per character, but was far more compatible worldwide, as it exactly duplicated the character codes from many existing sets. Unicode required fewer bits per character, but it lacked compatibility with character sets other than the ASCII set (on which it was, in many ways, modeled). Thus, the Unicode consortium, ostensibly formed to develop a universal code, merely raised its cultural bias to a greater level by proposing a code favoring ASCII (American Standard Code for Information Interchange) and all the hardware and software that used it; a 32-bit code may be more culturally inclusive, but the 16-bit Unicode uses less memory. Economical representation strikes again!

In 1992, after the ISO committee had bogged down in politics, the two standards were merged, with Unicode present as the first set of characters in the ISO 10646 (just as ASCII is the first set of characters within the Unicode set). The advantage of having the ASCII characters as the "first set" (that is, the characters encoded at the beginning of the numbering scheme) is that they can be represented in fewer bits, since most of the end bits are zeroes. Here again, by being in the very front of the series, ASCII characters are the "shortest" characters requiring the least amount of memory; all the ASCII characters fit in eight bits, whereas the rest of the Unicode characters fit in sixteen (and any characters added to the ISO standard outside of the Unicode set would require more than sixteen). So the merger, was, really, a victory for

Unicode, since it still allowed the ASCII characters to occupy the best position with the "shortest" characters, and also maintained compatibility with all previous ASCII files.[29] As yet, no characters have been assigned outside the Unicode set, and even the non-ASCII Unicode characters have yet to come into general usage. Thus, the current dominance of Western standards makes them the most compatible, and compatibility becomes the justification for their continuance. But this may change as foreign standards come about. As reported in *Wired*, "Bill Gates wants China's Ministry of Electronics to support Windows as a Chinese standard, but the Ministry has refused to do so. This has Bill hopping mad. "Microsoft Windows is known around the world," he huffed to the NYT [*The New York Times*] after talks failed with Beijing."[30]

Economical representation —using the least amount of memory possible, running faster, and in the end costing less money— is a major force shaping digital media, and the cultural materials digitized into it. This force does not necessarily always manifest itself as a deliberate bias; it is often merely present as a limitation. Although we have so far considered mainly images and text, anything translated into digital form taking up bytes of computer memory is affected. One case involving the spoken word, for example, is Marsha Kinder's *Blood Cinema* CD-ROM developed from her book of the same name. The CD-ROM contains printed text and spoken commentaries which are available to the user in both English and Spanish. Because Spanish is often more "wordy" than English, it often resulted in larger files, and Spanish versions of blocks of text appearing onscreen generally took up more room than their English counterparts, sometimes requiring a smaller font to be used. And, according to Charles Tashiro, the project's producer and designer, the audio clips in Spanish took up as much as 50 percent more memory than the same clips in English.[31] Given a limited amount of memory, then, the same materials may fit when in one language, and not fit when in another; not a deliberate bias on anyone's part, but when commercial value and the cost of production is taken into account, differences in size can become deciding factors in the setting of prices and even in deciding what gets cut or not made at all. Storage units with fixed amounts of memory like CDs will generally cost the same regardless of whether all the memory available on it is used or not, but producers will usually be trying to fit as much as possible into the memory available (which would probably still be the case no matter how much memory there was).

Thus far we have considered some of the biases occurring when analog media are digitized into digital media, and have yet to mention those works of culture that originate in digital media (which are obviously better suited for digital form), and which in many ways, are beginning to constitute their own forms of culture.

Digital Culture

The works produced on and around digital media can be seen as constituting their own subculture; one need only look at the works of authors like William Gibson or Bruce Sterling; the books in the "cyberpunk" genre (or indeed, anything using "cyber-" as a prefix); or any issue of *Wired* magazine, to see that a distinct kind of culture has arisen. In fact, there are now many subcultures within what we might term "digital culture", such as the cyberpunk movement, virtual reality enthusiast groups, and of course, the community of the Internet (and various networks and newsgroups within it), with all the jargon, legends, and innumerable subcultures that make it up.

There are many links and overlaps with other cultures coming together here; everything from hippie ideology of the 1960s, which sees virtual reality as a mind-altering technology and surrounds it with rhetoric once reserved for hallucinogenic drugs, to the punk movements of the 1970s, to 1980s corporate culture of the computer industry. What joins and holds these disparate elements together is more than a love of the technology and the production and use of media products; it is an outlook that envisions a vastly different near future in which the widespread benefits of digital technology will create a kind of utopia where human and machine will coexist (if, according to some, the distinction between them is not lost altogether). Its politics are a politics of information, which it sees as the all-important currency, the spread of which will help bring about the conditions needed for the utopia to exist. There is a radical quality to it which often paints the nontechnological as regressive, and considers technological illiteracy as backward. Digital culture's values are centered around technology and the application of technology to the problems of the world.

Developing at a time when the average consumer was able to become a producer and media author, digital culture differs from other past cultures in that the technological means of production are so central to the works produced on them that they are often the subject matter as well. A glance through any issue of *Wired* will verify this; not only do the layout and design of the magazine constantly remind the reader of its digital origins, but the entire content —from the main articles which center on new technologies and people using or creating them, to columns like "Fetish" (which describes itself as "Techno-lust unbound") and reviews of new products (pictures and prices included), right down to the advertisements which are almost exclusively all for computer hardware, software, games, conventions, and publications. The central place given technology implies a complete erasure of the boundary between "art" and the means of production or media on which it is made; hardware as well as software is in the spotlight, and someone like Bill Gates can become the subject of controversy, biographies, and admiration. The creation of technology is held on a par with the creation of art, and revered as such. Digital culture is not only aware of its origins, it is obsessed with them, and views everything else in relation to them. Through technology, it strives to be the origin of all art, or at least the means by which art is distributed and preserved. Technology is valorized above everything else, which is seen only in relation to technology.

Digital culture and its effects can even spill over into other areas of culture, for example, in metaphors of the brain as a computer, complete with "memory", ingrained in everyday speech. Noting how computers are changing language, Edward R. Swart writes;

> The interaction between computers and language is a two-way street. Although some grammarians undoubtedly abhor the use of the word *interface* as a verb, nowadays it's quite common to hear that people who don't get on well with each other don't interface as they should. Words that have languished because of neglect sometimes return with renewed vigor, and people talk of "the default option" in contexts that have nothing to do with computers. Phrases concocted by computer buffs have come to serve a generally useful purpose.
>
> Even a language like French, which resist the incorporation of English words and phrases, speaks of *les applications batch* and *le semiconductor*. The extent of interaction between English and world of computers can be gauged to some degree by the *Oxford Dictionary of New Words*. It contains detailed explanations of some 200 scientific and technical terms that have entered the popular vocabulary over the last decade, the vast majority from the world of computers.[32]

The effects on language indicate how computer-related concepts are entering into public consciousness, and how digital culture is being integrated more and more into everyday life.

Graphic design is another area that has certainly been affected by the flexibility afforded by media authoring programs in which various fonts can be combined (or designed), and the graphic properties of words can be emphasized, carrying on artistic explorations begun in the Bauhaus, now enhanced by the computer's capabilities and immediacy of results. In digital visual design, text and different forms of images are freely distorted, manipulated, and combined. Nearly every image is a composite one, made from a variety of component parts or even cultures. Digital culture's capacity for seamless compositing has encouraged production of composite imagery, the subject of the next chapter.

In digital culture, there is a tendency towards the absorption and assimilation of other cultures into digital forms, where they become raw material for image processing or sound bites or samples, ready for reuse in reappropriationist-style art. By providing a means —and a currency— of exchange, digital technology can help to spread cultures and encourage their intermixing; but to do so, all cultural materials must first be filtered through the grid of digitization, and then represented in as economic a form as possible in the limited memories of the storage or transmission devices where they will reside. Whether or not such a potentially homogenizing effect is desirable, cultural biases will exist, the greatest of them favoring digital culture itself.

NOTES

1. See Toscas, George J., "How Much Resolution is Enough?", *ON Production and Postproduction*, September 1994, pages 24-28. Similar discussions have elsewhere occurred concerning the number of samples per second needed for sound recording and reproduction. Although the sampling rate of the compact disc is high enough to make it probably the least-biased digital medium, there are those who feel the standard is still too low. Musician Neil Young is among them;

> Then there is Neil Young. ... The quality of digital sound, he said —the kind of sound stored on CDs— is "probably the biggest catastrophe that has happened to the recording of music in the last ten years or so," Young lashed out in a recent issue of CD Review.
>
> For him, the main problem is a lack of nuance. Robbed of such subtleties, digital fails to simulate the mind and the heart, and is bereft of "therapeutic value," Young claimed.

From "Vinyl/Digital sound isn't universally praised", *The Milwaukee Journal*, July 14, 1992, page D5. At any rate, the CD is less biased than early digital music's low sampling rate; writing about the recordable player-piano, Arthur Ord-Hume suggested that, "... some music lends itself much more readily to recording by a particular system than others." From Ord-Hume, Arthur W. J. G., *Pianola: The History of the Self-Playing Piano*, London: George Allen and Unwin, Publishers, ©1984, page 268).

2. Tanizaki, Junichiro, *In Praise of Shadows*, Translated by Thomas J. Harper and Edward G. Seidensticker, New Haven, Connecticut: Leete's Island Books, ©1977, page 9.

3. On olfactory symbolism, Constance Classen writes,

> The use of olfactory symbolism as a means of expressing and regulating cultural identity and difference is found in a great many societies. A particularly well-elaborated example of the olfactory classification of different social groups is provided by the Tukano-speaking tribes of the Colombian Amazon. According to this Amazonian culture, all members of a tribe share the same general body odour which is said to mark the territory of the tribe in the same way that animals mark their territories through odour. This territorial odour is called *mashá sëríri*, and has the metaphorical meaning of 'sympathy' or 'tribal feeling'.

From Constance Classen, *Worlds of Sense: Exploring the Senses in History and across Cultures*, London and New York: Routledge, ©1993, page 81.
4. Lammers, David, "Japan's gender gap", *Electronic Engineering Times*, Number 798, May 23, 1994, page 25. See also Tanizaki, Junichiro, *In Praise of Shadows*, Translated by Thomas J. Harper and Edward G. Seidensticker, New Haven, Connecticut: Leete's Island Books, ©1977, pages 1-8, where he describes how Western technology is out of place in the traditional Japanese home.
5. For some of the limitations and even dangers of using computers in the classroom, see Roszak, Theodore, *The Cult of Information: A Neo-Luddite Treatise on High-Tech, Artificial Intelligence, and the True Art of Thinking*, Berkeley and Los Angeles, California: University of California Press, ©1994.
6. Levinson, Paul, "The Amish Get Wired. The Amish?", *Wired*, December 1993, page 124.
7. Foster, George M., *Traditional Societies and Technological Change*, New York and London: Harper & Row, Publishers, ©1973, page 181.
8. See Rheingold, Howard, *The Virtual Community: Homesteading on the Electronic Frontier*, New York: HarperCollins Publishers, ©1993. On Japan, see pages 197-219; on England, see pages 235-240. On gender roles in Japan, see Lammers, David, "Japan's gender gap", *Electronic Engineering Times*, Number 798, May 23, 1994, page 25.
9. Greenwald, Jeff, "Dish-Wallahs", *Wired*, May/June 1993, pages 75 and 107.
10. Ibid., page 107.
11. Franklin, Jerry, "Of Whaling and Trigonometry", *Wired*, April 1994, page 32.
12. Since spatial and color resolution compete for computer memory, a balance must be struck in regard to the memory allotted each. Research has shown that for gray-scale images, spatial resolution is more important than dynamic range for maintaining image quality, while the reverse is true for color images. See Michael Ester, "Image Quality and Viewer Perception", *Leonardo*, Supplemental Issue 1990, page 51-63.

13. See Simon, Barry, "How Lossy Compression Shrinks Image Files", *PC Magazine*, July 1993, page 371-382; and Stan Miastkowski, "The Great Video Squeeze", *PC Graphics and Video*, March 1995, page 26.
14. Mike King, "Towards an Integrated Computer Art System", in Lansdown and Earnshaw's *Computers in Art, Design, and Animation*, pages 44-45.
15. Miastkowski, Stan, "The Great Video Squeeze", *PC Graphics and Video*, March 1995, page 24. Miastkowski's article also reviews and compares a number of commercially available compression programs.
16. Sibley, E. H., "Alphabets & Languages", *Communications of the ACM*, May 1990, pages 489-490.
17. Tayli, Murat, and Abdulla I. Al-Salamah, "Building Bilingual Microcomputer Systems", *Communications of the ACM*, May 1990, pages 496-497.
18. See "Bilingual Computer Runs IBM Software", *Byte*, March 1985, page 436; and "Toshiba Signs a Print Contract in China", *Electronics Weekly*, March 27, 1985, page 13.
19. See N. P. Archer, M. W. L. Chan, S. J. Huang and R. T. Liu, "A Chinese-English Microcomputer System", *Communications of the ACM*, August 1988, page 982.
20. Smith, Ben, "Around the World in Text Displays", *Byte*, May 1990, page 268.
21. For a diagram of the layout, see Figure 3. on page 30 of Joseph D. Becker, "Typing Chinese, Japanese, and Korean", *Computer*, January 1985, pages 27-34. Typing speeds are slower, but since characters represent whole words, fewer keystrokes are needed;

> According to character counts for identical documents, the number of English characters is 3.7 times that of Chinese characters. Therefore an input speed of 34 to 68 Chinese characters/minute would be equal to the 125-150 range for Western characters (25-50 words.)

From Cui, Wei, "Evaluation of Chinese Character Keyboards", *Computer*, January 1985, page 55.
For other methods of inputting characters see Lammers, David, "Whither Goest TRON? (The Realtime Operating system Nucleus project)", *Electronic Engineering Times*, February 20, 1988, page 22. For examples of these methods, see Jinan Qiao, Yizheng Qiao, and Sanzheng Qiao, "Six-Digit Coding Method", *Communications of the ACM*, May 1990, pages 491-494; and Jack Kai-tung Huang, "The Input and Output of Chinese and Japanese Characters", *Computer*, January 1985, pages 18-24.
22. Cui, Wei, "Evaluation of Chinese Character Keyboards", *Computer*, January 1985, page 54.
23. Becker, Joseph D., "Typing Chinese, Japanese, and Korean", *Computer*, January 1985, pages 31-32.
24. Jian Sheng explains;

> The initial consonants of Chinese characters are: b, c, d, f, g, h, j, k, l, m, n, p, q, r, s, t, w, x, y, z, ch, sh, and zh. The endings of Chinese characters are: a, an, ang, ao, ai, e, en, eng, ei, er, i, ia, ian, iang, iao, in, ing, ie, iu, iong, o, ong, ou, ua, uai, uan, uang, ue, ui, un, uo, ü, and üe. (Note that the initial and final elements shown here are those that appear in the standard spelling form and contain no phonetic information.)
>
> Chinese syllables take the form of either *initial-final* or *final*. The appearance frequency of the latter type is very low. (Even the final "an," which has the highest appearance frequency, places 181st in overall syllable frequency.) Therefore, most syllables comprise an initial-final pair.

From Sheng, Jian, "A Pinyin Keyboard for Inputting Chinese Characters", *Computer*, January 1985, page 61.

25. Skjellum, A., "Making Languages English Independent", *Dr. Dobbs Journal*, September 1984, pages 8-10.

26. Becker, Joseph D., "Typing Chinese, Japanese, and Korean", *Computer*, January 1985, page 29.

27. See "They Cannot Be Ignored", *Asian Computer Monthly*, April 1985, page 20.

28. Negroponte, Nicholas, "The Fax of Life: Playing a Bit Part", *Wired*, April 1994, page 134.

29. On the story of Unicode, see Frank Hayes' column "Consortium forms universal character code standards", *UNIX World*, May 1991, pages 109-110; and Dvorak, John C., "Kiss your ASCII goodbye (The Unicode standard supports foreign language character sets and replaces ASCII)", *PC Magazine*, September 15, 1992, page 93.

A friend of mine who worked at Sun Microsystems in the Bay Area said that there was a bias against making characters different sizes because it would make the English letters look smaller than the others, and nobody wanted to look like they were being biased towards English. He referred to this as "a PC [politically correct] movement that caused bad programming", with an overall negative effect, because it resulted in inefficient programming which slowed programs down a great deal, since they had to check if characters were 8-bit or 16-bit instead of all being one or the other. However, this notion of "speed", that everything should run as fast as possible, can *itself* be seen as an ingrained Western (and particularly American) value which is not always found in other cultures.

30. From the "Electric Word" column in *Wired*, June 1994, pages 28-29.

31. "...the Spanish files were considerably larger--almost 50 per cent greater. Of course, that's not purely a result of language difference, but of delivery, etc. But you can safely say they're bigger." From electronic mail correspondence from Charles Tashiro, received Wednesday, October 12, 1994.

32. Swart, Edward R., ""How Are You at Interfacing?" You may like it or not, but computers are changing the English language", *Byte*, December 1993, page 302.

5.
The Composite Image

At first glance, *Who Framed Roger Rabbit* (1988), *Citizen Kane* (1941), *Prospero's Books* (1991), the films of George Méliès, Home Shopping Channels, CD-ROMs, and the evening news may seem to have little in common. What they all share is the use of composite imagery —images and parts of images combined with text to create the finished pictures seen on-screen. In some cases, compositing is used to combine imagery into what appears to be a single, continuous, coherent space; *Who Framed Roger Rabbit* and many of Méliès' films use it to depict fantastic stories, while *Citizen Kane* uses it as an "invisible" effect, effacing its presence as much as possible. In other cases, compositing is not designed to create the appearance of a single unified space, but rather multiple imagery; either for formal experimentation as in *Prospero's Books* or *Powaqqatsi* (1988); for informational purposes as on Home Shopping Channels and the evening news; or as an interface in interactive media. Digital media's flexibility has greatly encouraged the use of composite imagery by making image manipulation easier, and by making it more available, both industrially and at the consumer level.

Of these two types of compositing, it is the former that will be given the most consideration here; the desire to create a seamless whole out of a collection of separately created elements has been a challenge pursued by filmmakers since the beginning of cinema and a main force driving the development of compositing technology. It is this area where the greatest degree of precision and preplanning is required, and one with a greater chance of failure —but also the most impressive when the illusion created is successful. Compositing used to create a unified space also tends to be more self-effacing and covert (as opposed to the overt nature of informational graphics), and is often more difficult to detect, making it a better vehicle for ideology as well. The focus, then, will be mainly on compositing used to create a unified space; and since considerations of sound and sound-image relations are beyond the scope of this chapter, "compositing" will refer strictly to the compositing of visual imagery, with most examples taken from film.[1]

By the time digital compositing technology appeared, the composite image was already very mature in form. Although forms of visual compositing first existed in painting and still photography, mechanical visual compositing began in 1825 when John Ayrton Paris invented the Thaumatrope, a cardboard disc on strings, which featured a picture on each side of the disc, for example a bird on one side and a cage on the other. When spun rapidly, the two images visually overlapped, due to persistence of vision, combining into an image of a bird in a cage. Printing two photographs on the same paper could also produce a composite, and photographic compositing was refined throughout the late 1800s. During the silent era, photographic compositing was creatively applied to motion pictures, flourishing in "trick" films like those of George Méliès and Ferdinand Zecca. As film technology developed, new techniques and greater control produced composites seamless enough to be used in a way that would not call attention to them, and they were soon employed for economic as well as aesthetic reasons; not only as tricks, but to save money, for example, by building models and photographing them instead of full-sized sets, and later compositing the actor into the scene.

As the number of films using compositing technology increased, the demand for better technology grew, and 1944 saw the appearance of the optical printer, a device combining a camera and a projector, in which film frames and sections of frames are rephotographed and combined, greatly increased the precision needed for optical line-ups.[2] It also introduced new processes and further systematized optical effects production. Since then, the optical printer has been continually refined and improved, culminating in the state-of-the-art techniques in films like *Return of the Jedi* (1983) and especially *Who Framed Roger Rabbit*. More recently, digital compositing technology, with its even greater potential for image manipulation, has overtaken most optical techniques. Today, in-camera, optical, and digital techniques all remain in use, each in their own niche, although most productions that can afford it tend to use digital compositing.

The creative possibilities of compositing have led to the film industry's increasing reliance on it for both small and big-budget films. In turn, the growing use of manipulated film imagery has changed the way films are made and received by audiences. In the following discussions of optical and digital compositing, a number of examples used in this chapter will be from the optically-composited *Who Framed Roger Rabbit* and the digitally-composited *Jurassic Park* (1993). These landmark films admirably demonstrate many strategies used in production and postproduction to accommodate compositing needs while meeting the demands of narrative and technical constraints. Once developed, these strategies feed back into the system and influence the continuing evolution of compositing techniques. A film's visuals and narrative are similarly affected, as is film form itself. Formal possibilities opened and encouraged by compositing further problematize existing notions of the *shot* and the *cut*, suggesting that these concepts be reexamined.

Who Framed Roger Rabbit and *Jurassic Park* exemplify these and other trends in compositing, such as the manipulation of discrete, overlapping layers of imagery involving techniques akin to those of cel animation. These methods require a fragmentary conceptualization of the image, which, although comprised of many constituent parts, is often designed to produce spatiotemporal continuity within the shot, and occasionally used to mask transitions between shots. Any number of aggregate parts may be entering and exiting the frame, or linked together in such a way so as to cause 'cuts' to be more spatial than temporal (in terms of the film frames, not the narrative), since the entire image is not changing simultaneously, unlike conventional cuts in which one shot follows another in time. Sequences like these, then, cannot be seen merely as a linear sequence of shots, but rather as an overlapping series of *elements*, hierarchically ordered and assembled, into a layered image whose composition changes over time as elements weave their way in and out of the image.

As the boundaries between shots blur, so do those between production and postproduction work. The onscreen combination of two or more images can occur at different points in the production or postproduction processes; several techniques can be done in-camera at the time of the original photography, while others are done (or rather, used to be done) through rephotography on an optical printer. Video images are composited electronically, and photographed images can be digitized and combined in a computer. While opening up vast creative possibilities of the medium, compositing has increased the importance of postproduction and the time it takes, and the demands of postproduction have placed more restrictions on the production process. Often the two occur simultaneously during the making of a film, running parallel, the postproduction processes barely maintaining their "post-" distinction. Although production and postproduction have always been closely interconnected, the use of computer animation along with digital compositing to create image elements for combination with photographic ones further blurs the distinction between them, signaling changes in the current mode of filmmaking used by large studios. Emphases on segmentation and manipulation, and the farming out of effects work to specialty houses can create further logistic problems, or even threaten the stylistic unity of the film; thus preproduction takes on greater importance and requires detailed conceptualization and communication of the film's content and design. A director or producer who is unfamiliar with such specialized filmmaking processes can easily lose control of a film through misjudgments or errors in estimation of what can be done.

Before examining the effect that compositing (and especially digital compositing) has on narrative and ideological structures within the film and theoretical concepts, let us first consider different types of compositing and their impact on visual structure, on which narrative and ideology depend.

In-camera Compositing

Before the advent of the optical printer, the creation of composite imagery was done with in-camera techniques, including foreground miniatures, glass paintings, front and rear screen projection, and multiple camera passes. Although in the film industry today the term "compositing" will usually refer to optical or digital compositing, in-camera techniques are similar in what they accomplish, and have some of the same effects on film imagery.

Several techniques involve placing objects close in front of the camera in such a way, so that from the camera's point of view, they appear to be part of the set in the background. Foreground miniatures are models made to scale and positioned in front of the camera so as to appear to be part of, or connected to, the larger and more distant live-action set. Glass paintings are used in a similar way, differing only in that they are flat images painted on glass sheets through which the actual live-action sets are seen. Mirror shots, involving semi-transparent mirrors or partially silvered mirrors, are also used to combine a live-action scene with a painting or model; the mirror is placed in a hole in the model, and is set to reflect the actor or scenery to make it appear as a part of the shot.

Although both of these techniques are very economical and relatively simple technically, they share several shortcomings: camera movement is severely limited,[3] lighting changes within the shot are difficult to match, and live-action elements are unable to interact with or even pass in front of the model or painting. Whereas foreground miniatures and glass paintings are static objects, front and rear projection systems allow moving imagery to be combined with live action. In both cases, the live action appears in front of a screen onto which a film is projected and used as the background of the shot (for front screen projection, a Scotchlite screen is used, and for rear screen projection, a Stewart screen is used). Although the process can be economical, camera moves are still rather restricted, image distortion due to off-axis projection can be a problem, and a method of synchronization is required between camera and projector. The projected footage limits the flexibility of the lighting, the set, and often the live action itself, while its grain, lower contrast range, and flatness can destroy the illusion of depth in the image. Many of these problems have been solved by newer technologies, and in-camera compositing remains in use today. The Introvision system, for example, has updated in-camera techniques and integrated them even more closely with production techniques, giving more direct control of shots to the director. An article in *American Cinematographer* explains the Introvision process and its appeal to directors;

> In recent years ... directors like Bruce Beresford, Sam Raimi, John Avildsen, Andrew Davis, Peter Weir and Barry Levinson have found a way to take back control over the visual effects in their own films through the relatively inexpensive and danger-free Introvision dual screen process. The director remains the maestro of the filmic orchestra, able to guide actors in a natural soundstage environment and comfortable in the knowledge that the eyepiece of the camera contains all that will appear in the final film. ... *Fearless*' harrowing passenger's eye view of an imploding plane interior was achieved in only three days of filming on a tiny shooting space in Introvision's Hollywood headquarters. ... A beam-splitter mirror which both reflects and transmits light acted as the nodal point in the process. Real-time VistaVision footage of a turbulent, fire-ravaged plane interior (a miniature set designed and shot by the Introvision team) and exterior scenes of a cornfield, which were to act as a backgrounds, were transmitted onto and through the mirror from a film projector placed to the side at a 45-degree angle.
>
> Director Weir and cinematographer Allen Daviau, ASC could then view this projected background footage on the mirror from the eyepiece of their camera, which was set at the same 45-degree angle and elevation as the projector, but placed on the other side of the beam-splitter. The actors and set props (sometimes only Jeff Bridges and a single plane seat) were then set up between the camera and the mirror, so that the filmmakers could shoot live action against the projected background. Filming additional takes simply required running back the projector and starting over.[4]

And, according to Introvision vice president Andrew Naud,

> Unlike digital compositing, where you composite a lot of the elements on a computer and put it all together over a long period of time, our process can be done instantly through the eyepiece of the camera... We've been doing this for ten years, so all the current buzz about digital is quite amusing to us.[5]

Even in an age of digital compositing, in-camera compositing has kept pace and carved out its own niche of what it does better than optical or digital methods.

Finally, the onscreen appearance of a video display, monitor, or film screen is another way of combining two images in-camera. In this case, however, the image appears diegetically *as an image*, not just what that image represents. This is not really compositing, since the displayed image is, as an image, already a part of the film in which it is contained, appearing as itself and not something else; but this imagery is still created separately and combined into the image through the use of film or video equipment, similar to front or rear screen projection. In some cases, however, like the video war game room in *Toys* (1993), the onscreen images *are* added in postproduction, despite their existence as diegetic images.

In most of the compositing methods mentioned above, the image is created so as to match the perspective of the camera's point of view, which records the image in a single pass. Often, however, it is not possible to construct the complete image all at once, and so more than one pass of the same piece of film is required. Multiple camera passes create superimpositions, but the individual passes are carefully designed so as to appear to be a single image (for example, a pass which adds the glow to a spaceship's engines will be shot in the dark, with only the glow appearing). Introduced around the turn of the century and still in use today, this method is similar to the optical printer but less versatile since the camera position must be precisely controlled so that elements shot on different passes will match up exactly. In-camera compositing of moving camera shots have become possible with the invention of motion control systems (first used for *Star Wars* (1977)), in which computer-controlled cameras duplicate moves exactly on each pass. By using several passes on the same strip of film, the images are composited directly onto the camera original, saving extra generations and maintaining image quality. On the other hand, multiple passes of the same strip of film increase the chance of scratches, worn sprocket holes or film jams, and should a mistake appear on any of the passes, the entire shot must be redone. Unlike the other methods discussed thus far, multiple passes are not made simultaneously; events or actors who are meant to interact with each other but who appear in different elements or passes must have precise timing, and often play to an empty space where other elements are imagined to be.

In the special effects branch of the film industry, *compositing* is usually used in a narrower sense to refer to optical work done in a film lab and digital compositing done in a computer. This distinction is important in defining a division of labor; compositing and special effects work are often farmed out to companies specializing in these processes. These companies shoot many of the effects elements as well, particularly when in-camera techniques are needed, and where technical concerns are beyond what a production crew would normally encounter.

Optical Compositing

Using no computer compositing, *Who Framed Roger Rabbit* (1988) took optical compositing far beyond what had been done before, winning an award from the British Academy of Film and Television Arts and the Academy Award for Best Visual Effects in the process. It broke the record for the largest number of process shots used in a film and for the most elements ever composited together within a single shot (the one in which Eddie Valiant walks through the Maroon studio backlot encountering a number of cartoon characters including a pelican on a bicycle and the brooms from *Fantasia*). According to Ed Jones, Director of Postproduction at Industrial Light and Magic, "There were 162 original elements. When you add in all of the mattes and intermediate stocks, we were probably working with 400 to 500 pieces of film. There were around 300 passes through the optical printer. And that's all for an illusion which lasted a couple of seconds."[6] Although *Who Framed Roger Rabbit* represented the cutting edge of compositing technology and a dazzling display of the optical printer's creative power, it ironically appeared at a turning point in compositing history when compositing began the transition from the optical printer to the digital technology of the computer. Thus, the film now stands as a monument to a dying technology, the pinnacle of its achievement.

On the optical printer, pieces of film are rephotographed onto a single negative, and images can be manipulated with special mattes and gates allowing the image to be panned, rotated, zoomed in or out, or cut into sections. Some fairly common uses of the optical printer are dissolves, wipes, push-offs (when one image appears to push the preceding one offscreen), split screens, superimpositions, and subtitles and film titles appearing over live action. While dissolves, wipes, and push-offs are used as transitional devices, split screens, superimpositions and subtitles usually denote simultaneity and demand that the viewer's attention be divided between multiple elements. Split screens generally contain different points of view or locations, calling attention to differences, while superimpositions, which overlap, invite comparisons or make associations between images. Subtitles, as a simultaneous translation of dialogue, tend to divert a viewer's attention to the bottom of the screen, where they most often appear. Unless they are in the original version of the film, they differ from other forms of compositing in that they are added by someone not overseen by the director, to finished films which (usually) have not been graphically planned to allow for their inclusion, resulting in the disruption of visual patterns and compositions which would otherwise lead the eye, as in a painting. Subtitles also emphasize the flatness of the screen, and, along with bad translations and misspellings, make the viewer conscious of both the viewing process and the cultural gap present between languages.[7] Unlike subtitles, a film's titles and credits are often placed over images at the beginning and end of a film. These titles are usually planned for, and do not occur constantly throughout the film, except perhaps to segment a film into chapters, or in some cases for an aesthetic effect; an example being the "infography" used by Peter Greenaway in *Prospero's Books* (1991) and *TV Dante* (1990) which appeared as an aesthetic element carefully integrated into the visual design. If these elements distract the viewer, the distraction is controlled and directed.

Apart from nondiegetic segmentation and punctuation, most optical work involves creating self-effacing (or at least integrated) "invisible" effects, wherein elements are combined into the same image so as to appear to be existing within the same space. The most common of these effects involve traveling mattes and matte paintings. In the matting process, different parts of the same scene (the background and the foreground, for instance) are shot separately, but with an attempt made to match camera angle, perspective and lighting. The filmstrip of the foreground element is then printed onto a high contrast stock, and a hold-out matte is made from the element. This will be placed in front of the background element and rephotographed, the black matte leaving portions of the film unexposed. This area matches the shape of the foreground element, which is later printed into the unexposed area on a subsequent pass. Although usually done on an optical printer, this process could be done with aerial-image printing on an animation stand.[8]

Matte paintings are similar to glass paintings, and usually remain fixed during a shot. Often, however, elements move during a shot and require traveling mattes, which change from frame to frame allowing the element to move. In either case, the matting process can be used very successfully to give the impression of a unified space with objects in it. Usually traveling mattes are used to move an object or character over a background, but the technique improved to the point where intricate kinds of interaction between characters or between a character and a background are possible, as in the scene in *Who Framed Roger Rabbit* in which Eddie tries to catch Roger and throw him out of his apartment after finding him in his bed. The high degree of accuracy achievable in the optical line-up has made this possible, allowing the space of the frame to be meticulously planned. Raymond Fielding's book, *The Technique of Special Effects Cinematography*, gives ±.0001 inch as the accuracy of an average optical line-up.[9] State-of-the-art technology like that of ILM is even more precise, and digital technology using film recorders which use lasers to print the image onto the film have registration as perfect as the filmstock used allows.

Because optical compositing works with layers of imagery, composited elements can be used to hide other objects. For example, in *Who Framed Roger Rabbit*, when Roger and other animated characters interact with the live-action environment, mechanical effects devices stand in for the composited character. The fact that the foreground elements of the animated characters overlap and cover the background means that these devices can be hidden behind them. For instance, a waterspout was used for the shot in which Roger pops up out of the sink and spits water, and an armature was used in the shot where Roger lifts a shotglass and tosses it over his shoulder. Devices covered by the animated characters are designed to be as unobtrusive as possible, but this can still put limitations on the animation which must cover the device and attempt to look "natural"; when Roger throws the shotglass, the movement of his arm is unnaturally stiff, due to the armature holding the glass; when he spits water out of the sink, his neck and lips must remain in place to cover the water pipe.

Occasionally, a person may even be hidden behind an element. Throughout the chase when Eddie rides Benny the Cab, Bob Hoskins was filmed riding a stripped-down auto chassis, the driver of which was hidden low in the back of the car. When the animated cab was drawn over the chassis, the driver's head, which was constantly in motion, often stuck out far from the back of the cab. To cover it, a rear tire was added to the design of the cab, which moved fluidly and covered the driver's head from frame to frame. Dale Baer, of Baer Animation, recalls the difficulties; "...the hard part was the driver in the back, because that's where the spare tire was, and you'd be going along fine, covering him up, and the driver would decide to move, to look around the car, and throw everything off; you had to go back in and readjust the spare tire, just because of that one little move he made. It's not like today, where you're going into a [computer] paintbox... you had to go back in, and redo everything that you'd already done."[10]

Mattes can also be used to produce lighting and shadow effects, allowing composited elements to appear to be interacting with the light in a scene. Although they must match the characters' movements frame by frame, these mattes are produced separately and often mean additional passes through an optical printer. *Who Framed Roger Rabbit* uses a great variety of these effects, including backlighting, edge and side lighting, highlights, moving light sources, and combinations of these. In several shots of the rot-gut room sequence where Roger hides from Judge Doom, Roger moves through a complex, continually changing pattern of lighting. Shadows fall on and are cast by the animated characters, and their reflections appear on shiny surfaces; examples of this can be seen all throughout the piano duel scene. Often, reflections need only be present in order to be convincing, rarely are they onscreen for long or the focus of attention. During the piano duel between Daffy and Donald Duck, Daffy is reflected in the black piano, but his reflection is sometimes hard-edged and specular, and at other times so soft and diffuse it is barely visible. Sometimes reflections do not even have the correct perspective. In one of the last shots of the opening inset cartoon, *Somethin's Cookin'*, Roger's reflection in the floor is incorrect in every frame.

The increasing number of elements and complexity of these shots make optical compositing a long process which can require an image to go through many generations, each decreasing the film's sharpness. Despite computer-controlled optical printers, film jump and weave, worn sprocket holes and human error inevitably degrade image quality. Digital compositing done in a computer avoids many of these problems; because images are stored in digital form, there is no generational loss, and images can be manipulated down to individual pixels. Computers offer more flexibility and greater control over image production and error correction, and their reduction in cost helped accelerate the changeover to digital compositing.

Digital Compositing

Digital compositing evolved out of analog electronic compositing, a process normally associated with video processing. The first analog electronic system was patented in 1932, and was used to combine images coming from two different cameras for broadcast television. Digital compositing combines electronic compositing with a computer, allowing for greater image resolution, manipulation, and storage. The image can be manipulated down to the individual pixel, and these techniques have taken over a number of optical techniques; for example, the matte paintings in Disney's *The Three Musketeers* (1993) were done digitally.[11] Digital compositing can do more than just remove matte lines and make color corrections; a much closer interaction between elements in a scene can exist, and between the environment and the composited object. For some outdoor shots in *Jurassic Park*, the

Tyrannosaurus Rex walks between objects in foreground and background, and interacts with the light and even with the rain, and in the Gallimimus herd scene, dinosaurs react to landscape, lighting, and actors, while moving in perspective matching that of the scene. Digital compositing combines some of the best elements of optical compositing with those available in animation.

Disney's animated features made after 1991 have been assembled in digital form, utilizing a computer paintbox system instead of conventional methods; all the pencil drawings are digitized into a computer, where they were "inked", "painted", and composited together. Here too the computer can do things that would take far too long by hand; for example, in *Beauty and the Beast* (1991), the rouge on Belle's cheeks appears to have been airbrushed in on every frame. On the pencil drawings for the film, small triangles were drawn on her cheeks where the rouge would be centered, and the paintbox program softened the edges around the area, blending it in with the rest of her face.

Elements are also created completely through the use of computer animation, and films like *The Abyss* (1989), *Jurassic Park: The Lost World* (1997), *Starship Troopers* (1997), *Titanic* (1997), *Star Wars Episode I: The Phantom Menace* (1999) and dozens of other films, show how these elements can be believably combined with live action. Animated films like *The Great Mouse Detective* (1986), *The Little Mermaid* (1989), *The Rescuers Down Under* (1990), *Beauty and the Beast* (1991), *Aladdin* (1992), *The Lion King* (1994), *Pocahontas* (1995), and *Hercules* (1997) combine computer-generated elements with hand-drawn animation in various ways. The flexibility and control available in the computer make compositing similar to animation, which its control over the creation and manipulation of imagery. Animation's methods of creating imagery are now available to live action as computer animation grows in its ability to simulate detailed three-dimensional objects and environments, resulting in the convergence of live-action filmmaking and animation.

Digital compositing also makes a 'cel layer' approach possible in live-action filmmaking. Cel layering in animation is the process in which the elements are painted onto clear sheets of celluloid and then laid over one another on top of a background. The animator controls all the elements and movements, and can even move objects or characters from one layer to another. An object may exist on a separate cel layer for a part of a shot, while a moving character appears on another, but when the character picks up the object, both can then appear on the same layer. Because this type of change can occur from one frame to the next, it can be difficult to separate one layer from another. In cel animation, color saturation differences due to the layering of the celluloid sheets and movement patterns (including timing and overlap) often give clues to the image's construction. As recent Disney films have shown, digital compositing is a good way to avoid many such problems.

This kind of layering can also be found in optically composited films; in *Who Framed Roger Rabbit*, especially in the Toontown sequence. For instance, in the shot where Eddie gets out of his car and looks up at Jessica's window, Eddie is a live-action foreground element on an animated background. Jessica's car, parked in that background, appears to be a live-action element as well, but it is actually a photograph cut and pasted to the background. This method was used because Eddie would never pass behind the car. Towards the end of the shot, as the pan to the window begins, the live-action element of Eddie is freeze-framed and, like a cut-out, moved out of frame. The "cel layer" approach is an efficient way of conceptualizing the composited image in live action, and digital compositing can now do things that would be far too costly using traditional animation methods.

Digital compositing has also been combined with in-camera methods, resulting in real-time compositing systems which combine live action with a low-resolution version of the computer imagery and provide feedback for the director, actors, and animators. Such a system was used for Jim Henson's Waldo C. Graphic, a computer-generated character (made by Pacific Data Images) capable of realtime interaction with the Muppets. And finally, as mentioned in chapter four, real-time composited feedback could be used with motion capture and physically based modeling to create scenes with computer-generated creatures and environments combined with live-action ones, with control of these elements available during the shooting stage.

Opening in 1992 in Los Angeles, Eastman Kodak's Cinesite, with its Cineon digital film system, was a milestone in digital compositing history. With a film scanner and recorder operating at a resolution of 166.67 lines per mm, the system can produce high resolution imagery indistinguishable from camera negative. Film can be scanned in, composited, and printed back onto film without any compromise of image quality. A single frame of 35mm film can take up to 40 megabytes of computer memory, and the cost of scanning and printing are still somewhat expensive.

One way to avoid scanning is to record digitally during shooting, which a growing number of projects are doing, although digital cameras do not yet have the kind of resolution that film does. For the short sequence in *Dreams* (1990) in which a character walks around inside a series of Van Gogh-like paintings, Akira Kurosawa used HDTV to shoot and composite images and then transferred the finished product to film. Peter Greenaway's *TV Dante* and his feature-length *Prospero's Books* were composited entirely on HDTV, the latter transferred to 35mm film and released in theaters.

Many systems like HDTV which can produce digital imagery from the recording phase onward still operate at a greatly reduced resolution compared to film, and film technology is still improving. According to documentation available from the company, Eastman Kodak's 5384 (numbered 7384 for 16mm gauge) color print film has a resolving power of 630 lines per mm with a test object contrast of 1000:1, and 250 lines per mm with a test object contrast of 1.6:1. Kodak has even sharper films than this, but this stock is a more commonly used one, at least for prints. Camera negative often has much fewer lines of resolution.

Digital cameras will likely catch up, aided by advances in micro- and nanotechnology, and shooting and scanning will be identical. Compositing can be done completely seamlessly, since computers allow image manipulation to occur down to the individual pixel; even the boundaries between elements can be eroded or smeared, morphing and blending elements together into a field of pixels.[12] However, commercial enterprises without the time or money required for this kind of detail and precision are willing to settle for something that looks reasonably good. Lower resolutions reduce postproduction costs and can still look stunning, as *Prospero's Books* demonstrates. But, like any other process, compositing has its limits.

Limitations and Spatiotemporal Defects

All digital images, because they are broken up into a grid of pixels, experience some degree of spatial *aliasing*, the noticeably jagged edges of low resolution computer graphics resulting from the breaking up of the images into pixels sharply defined from one another. High resolution graphics are much smoother in appearance, but the larger number of pixels requires more time for image manipulation and processing; thus higher resolutions require higher budgets.

Composites almost always need some cleaning up or color adjusting for several reasons. Spatially, the film must try to represent a three-dimensional scene within a two-dimensional plane, with a limited amount of resolution. Even with fine-grain photography, compositing can often have side effects which destroy the illusion of space by calling attention to the image's flatness. If a composited element's lighting is not matched to the onscreen source or background lighting, or if its colors are timed differently, it will stand out from the rest of the image and the conflicting spatial cues and information will cause the image to be read as flat or not in the same space. *In the Line of Fire* (1993) contained shots in which characters are composited into actual campaign footage shot prior to the film; the campaign footage has a soft, desaturated look while the characters composited into them look sharp and slick, and without believable outdoor lighting, they end up looking pasted into the shot. Similar effects were used in *Forrest Gump* (1994) more seamlessly, although graininess is noticeable in several shots.

The same can be said of incorrect matchings of focus, perspective, and camera movement, and the incorrect overlapping of elements. In shots with large numbers of elements, layering errors can occur, as in a space battle shot (known as SB-19) from *Return of the Jedi* (1983) where two tiny and supposedly distant TIE fighters pass *in front of* the large, forefront element of the Millennium Falcon for a few frames. Likewise, matte lines appearing around the edges of an element separate it from the background and also tend to emphasize the flatness of the image. In some cases where computer animation is used, an entire shot will often be digitized so that the resolution of the computer graphics matches that of the live action; the elements will then match within the shot, but the digitized shots themselves can appear slightly different from the analog ones, in resolution, contrast, and occasionally color saturation. An extreme case of this appears in the rather low resolution digital shots in *Drop Dead Fred* (1991) where process shots containing computer graphics are juxtaposed with straightforward photographic ones; the difference in granularity is very noticeable. This effect is not only distracting, but allows the audience to anticipate effects in a shot before they occur. New digital film systems have overcome problems of resolution and color reproduction, although not all production companies can afford these state-of-the-art systems. Nor is all the work done on these systems that could be to correct and smooth out the image, because of the long hours and high costs involved. Even with digital techniques, differences in granularity in the resolution of the individual elements scanned in can occasionally remain visible.

Temporal aliasing, or strobing, occurs when elements move but are not motion-blurred, resulting in images that remain sharp-edged and seem to move by discrete amounts. The faster the movement, the greater the displacement between successive images, until the Gestalt is lost and the motion no longer seems continuous. This effect is particularly noticeable when the foreground element is moving quickly, or when the background is brighter than the element traveling in front of it. (Strobing of this sort can be clearly seen in *The Rocketeer* (1991) in the shot where characters swing in front of a burning blimp, and in *Return of the Jedi* (1983), in the shot where the desert skiff flies in front of the exploding sail barge; almost any film in which some character or vehicle flies past an explosion will contain strobing.)

If a soft edge is used, however, combined with slow, smooth movement, the effect can be quite convincing. For example, in one shot from *Who Framed Roger Rabbit*, Roger sits in the dark, watching an old Goofy cartoon on a bright movie screen. The screen is in sharp focus while Roger is in soft focus, his ears slowly undulating, and the shot gives the appearance of having depth. Another way of avoiding strobing in hand-drawn animation is the fluid stretching or repeating of objects as they move, so as to fill the area normally occupied by a blur. By lengthening the objects so that successive images share some overlap, motion appears to be fluid from one point to another, and no 'jump' is apparent. An example of both techniques occurs in the shot where Lena Hyena runs across her room to Eddie Valiant. Not only does she stretch out, but her eyes are repeated to create a stronger Gestalt of movement; in one frame she has five eyes in a row. The motion blur occurring in live-action film can be simulated with digital compositing; using various algorithms, it can smoothly composite any two elements, as long as the desired effect can be adequately visualized and be done practically.

Yet while technology can enhance a film's visual believability, it may contribute little to its plausibility, and technically perfect effects cannot always hide the composite nature of the imagery. Roger Rabbit and the other Toons are cleanly and believably composited into the three-dimensional space of the live action, but the very fact they are animated leaves no doubt that they are separately created and composited elements. The same can be said for flying spaceships, miniature people interacting with normal sized ones, people appearing with doubles of themselves,[13] ethereal ghosts, or oversize monsters. No matter how good the effects are, such things require a suspension of disbelief from the audience. Good compositing, however, does aid visual plausibility, making the suspension of disbelief much easier and keeping the spectator from getting distracted by the effects.

Besides good compositing techniques, the design of an element can help to believably integrate it into a scene. In both *Flight of the Navigator* and *Terminator 2*, computer-generated objects have chrome surfaces which contain the reflections of other objects in the scene, and in *The Abyss*, the water tentacle is clear and refracts light from objects behind it. And in all three of these films as well as *Who Framed Roger Rabbit*, *Death Becomes Her* (1992), *Jurassic Park* and many others, the composited elements happen to be characters in the film; while on the one hand this may require a further suspension of disbelief, the audience will at least consider the element as a character and not think of it merely as an effect. Even the water tentacle in *The Abyss* takes on a human face and interacts with a human character.

To aid in the suspension of disbelief and make these composited characters seem more real (as opposed to effects capable of anything), each has a set of ontological rules which it must follow. The water tentacle falls apart once its base is broken, the T-1000 terminator can only form certain shapes and cannot form complex machines, and according to Robert Zemeckis, in *Roger Rabbit*, "It came down to a simple rule, which was, a human could be killed by a toon gun, but a toon could not be killed by a human gun. That's sort of the rule we established and that sort of... trickled into all aspects of human/toon props."[14] By defining a set of rules and consistently adhering to them, the composited characters seem to be closer to humans whose behavior is also subject to physical laws and restrictions. Such rules guiding expectations often apply to vehicles or objects as well, and may require the narrative to work around them.

Theoretical and Narrative Implications

Compositing opens up a wide range of creative possibilities, and often brings success at the box office[15] where special effects are as much of an audience draw as sex or violence seem to be. The technical virtuosity of *Who Framed Roger Rabbit*, *Return of the Jedi*, *Jurassic Park*, *Forrest Gump*, or *Titanic* is a source of spectacle, akin to a magic show, and frequently such films gain further publicity through the airing of "The Making of..." documentaries revealing effects techniques. The documentaries and the visual complexity of the films themselves encourage return viewings, and can account for a large portion of a film's revenues.

Apart from its use as magic or spectacle, compositing can be used for financial reasons, when something is too large or expensive to construct full scale; for events too dangerous, improbable, impossible, or difficult to control; and for shots where an object or character is not to the proper scale, nonexistent, unavailable,[16] or appearing with multiples of itself. Although these instances can occur in any type of film, they are most likely to appear in genres which portray the fantastic, like science fiction (depicting advanced technology), fantasy (depicting magic), horror (depicting the supernatural) or action/adventure with its constant dangers. In non-fiction and avant-garde film, compositing, when it appears, is used mostly for formal effects with ideological implications, as in Godfrey Reggio's *Powaqqatsi* (1988) during the "Video Dream" sequence, or Pat O'Neill's *Water and Power* (1991). Straightforward dramas, comedies, or adapted stage plays usually have less need for compositing, though there is an increasing use of it for invisible effects (an early example of this, *Citizen Kane* (1941) featured a great deal of optical compositing, for compositions and sets that would have been too big or expensive).

Compositing and state-of-the-art effects, then, can influence the narratives of the films, affecting plots, characters, and locations. Characters utilizing composite imagery often tend towards the fantastic and possess qualities or abilities beyond those of a human, or not even be human at all. Taken to an extreme, this can affect identification, become distancing, or even destroy plausibility, a topic which shall be taken up later.

The narrative is also affected by compositing on the levels of preproduction, production, and postproduction. The great precision and control required by compositing necessitates comprehensive preplanning and design of the film's graphic content, resulting in the production of storyboards, decisions made concerning shot size and camera angle, and often a plot designed around effects sequences, one of which usually occurs during the film's climax. Preproduction structures and places greater restrictions on production; directors and producers may need to collaborate with effects technicians who often direct effects shots themselves in separate studio spaces, without the film's director present. Although it is the director who decides what shots are made, effects houses will often shoot painstaking shots involving miniatures, stop-motion, bluescreen elements, or explosions and so on at their own studios, and separate position titles, such as visual effects art director or supervising visual effects editor, are listed in the credits.

Before shooting *Who Framed Roger Rabbit*, Robert Zemeckis conferred with Richard Williams, whose team would animate much of the film. Williams agreed with Zemeckis that camera moves should be used (despite the problems they would create for the animators), instead of the static camera that had been used in the past for animation/live-action combination films. However, this division of responsibilities did, on occasion, create conflicting interests regarding the design of shots in the film. In *Roger Rabbit*, for example, there was a dispute between Robert Zemeckis and the animators over a shot in the opening cartoon. As Zemeckis recalled;

> There was one large argument over the speed at which the chili bottle, or the hot sauce bottle, slides down the shelf... an animation shot like that probably hasn't been produced in thirty years, the reason for that is because it's [the bottle's] writing that is shifting perspective in absolutely every frame. It's a lot of words in a form coming straight at the camera... nobody has the money to pay for the manpower to do a shot like that anymore, because it took incredible mathematics and

> grids and man-hours and whatever to do this. So, naturally, the animators, having finally a chance in their life to do a piece of film like that, wanted it to be long... So they had the shot twice as long. And I got my hands on the film, and I basically cut it in half, and of course they went crazy and they said, don't you realize how rare a piece of film like this is? I said, I don't care, it's a stall, the timing isn't right. So the compromise was that we step-printed it...[17]

In this case as in others, it is the director who makes the final decision, which in turn will reflect his or her own knowledge of effects technology. Although the director or producer may have the final say, they are increasingly reliant on effects technicians for an idea of what can or can't be done, and what the cost will be.

Other members of the crew, such as the cinematographer, camera operator, and the sound team are also affected by compositing. Lighting has to match and compensate for elements not present, the camera must allot space for them in compositions, and frequently the sound crew will have to design sounds for nonexistent objects such as science fiction hardware or vehicles, creatures, or supernatural phenomena. But perhaps most affected are the actors appearing in the film. Besides concentrating on the scene at hand, they must appear to interact with nonexistent objects or characters, moving through designated areas and reacting with precise timing. Even the set itself can be absent, when the actors are shot as separate elements alone in front of a bluescreen, where they are especially dependent on the director's cues and their own conception of what the shot is supposed to look like.[18] The methods used to shoot bluescreen shots of actors are the same as those used for spaceship models or other objects, and occasionally, even movements of the actor are simulated through camera movements; when Eddie Valiant falls after losing his grip on the flagpole in Toontown, his fall was simulated by filming the shot sideways and pulling the camera back. In the side view shots of Eddie falling past the building, he falls at a uniform rate without experiencing the acceleration he would have under normal circumstances.

Actors composited into a shot lack presence within the depicted space during production. Afterward, they are at the mercy of technicians who can alter any aspect of their images, put multiples of them onscreen simultaneously, or even stretch and distort the images spatially or temporally. As photographic elements, the actors can be reduced to the status of mere onscreen graphics, arranged in flat, overlapping layers which shuffle the elements around, display them, and discard them a moment later. Even the creation of the elements does not depend on the actors, who can be simulated through the use of miniatures, dummies, puppets, or animation. For a dangerous shot in the car chase with

Benny the cab, Eddie is simulated by hand-drawn animation, and by distorted still images in several shots in the Toontown sequence; this can be seen quite clearly by examining the individual frames in the shot where he appears to fly upwards across the screen when the elevator comes to an abrupt stop. Usually, shots which contain a visual surrogate of the actor appear only briefly, with the surrogate partially hidden or in motion, and some distance away from the camera; all strategies are deliberately designed to obscure detail. For larger budget productions, however, these methods are less necessary now that computer animation can create three dimensional models which can stand up to scrutiny, even in a closeup, and remain onscreen longer; good examples of this are the *Jurassic Park* dinosaurs or the feather floating on the wind in the opening shot of *Forrest Gump*.

Computer animation and digital compositing blur the distinction between production and postproduction even further. Rendering algorithms can produce computer-generated elements indistinguishable from their photographic equivalents, if they even have them. Similarly, models, puppets, and explosions have referents but are shot with temporal and spatial cues that deliberately mislead the audience. Both of these types of shots are usually created after principal photography; as growing emphasis is placed on postproduction, it begins to overlap more and more with production.

As the technical overtakes the aesthetic, it may even come to depend on it; one danger of advanced postproduction techniques is a "fix it in post" mentality, leaving problems encountered in production to be solved or 'cleaned up' by the technical wizards in postproduction, whose manipulatory powers over the image are assumed to be all-encompassing.[19] Along with the piecing together of the film they already do, editors may be asked to hide mistakes with fast cutting or through cutaways; and, due to storyboarding, there may be less coverage to work with. The fast cutting found in many effects sequences may produce the desired Gestalt of matching action, but a frame-by-frame scrutiny can sometimes reveal shortcomings of a technical (and narrative) nature, where continuity wavers and weakens, seams loosen, and the film reveals the marks of its creation. Run at normal speed, however, the audience barely has enough time to keep up with the narrative, much less reflect on its construction.

Besides the reading of the image as a unified, interactive, three-dimensional space, films often require a great suspension of disbelief, with occasional leaps of logic or plausibility. Good technique and state-of-the-art technology can provide the desired representation of space and interaction, but only the narrative can supply plausibility, which often is influenced more by genre conventions and expectations than by any congruence to actual lived experience. Such films follow their own internal rules of behavior, limiting characters' powers, technological abilities, magic, and so on; how closely such rules are followed and how reasonable or complete they are to begin with will be a factor determining their plausibility and success. And the quality and seamlessness of the visuals will still remain an important factor in any suspension of disbelief; even the most clever magic trick can fail if poorly executed by the magician.

The question of plausibility leads to ideological implications of compositing. As an advanced technology itself, it is not surprising that compositing usually endorses technology; even films which are supposedly anti-technology, like *Terminator 2*, will often dazzle audiences with special effects (which may have attracted the audience in the first place).[20] This encourages belief in the idea of limitless technological growth and advancement, in an 'anything is possible' universe where the laws of physics are routinely broken and inconsistencies are elided through aesthetic means; if it looks possible, it is possible; if it can be shown, it can happen or exist. Other developing technologies in the image industry have taken the same stance, carrying it even further; virtual reality (whose very name implies a negligible difference from lived experience) would have us believe that not only can we watch computer generated worlds, but we can enter them, touch things and interact, and even see ourselves present inside them.[21]

Due to its very nature, compositing can become a dialectical activity; just as Eisenstein showed how montage can be used dialectically, similar activities can occur between elements within the frame (which Eisenstein mentions in his discussion of composition). Compositing can be used to promote hybridization, plurality, and dialogism, and its fragmenting and restructuring of the image make it an analog of postmodern texts in which aggregate parts are combined into a whole. On the other hand, these parts are not necessarily homogeneous, particularly in the realm of the fiction film, even though they are being used to create an illusion of seamlessness. (Although the elements in the fiction film tend to be designed so as to fit together seamlessly, the compositing present on news, sports, weather, home shopping channels, music videos and avant-garde films tends to be less self-effacing, since it is done for informational or nonnarrative purposes, often using text and graphics in their compositions.) Shots that are made up of a large number of elements, whether heterogeneous or not, are usually used as spectacle and tend to be decentered compositionally, allowing the shot's complexity to take full effect, with a majority, if not all, of its parts in motion. Decentering and hybridization occur on higher levels as well; films like *Blade Runner* (1982), *Who Framed Roger Rabbit*, *Back to the Future III* (1991), *Star Wars*, *Young Sherlock Holmes* (1985), *Titanic* (1997), and many others hybridize two or more genres together, at least one of which is a genre favoring the use of compositing. Compositing allows genres to interact, while favoring the technological, magical, supernatural, or exaggerated in the storyline, especially in sequences where the film diverges most into the fantastic.

Hierarchies of Elements

Compositing may promote hybridization, but one in which components do not share equal status. In any film containing multiple elements, there is some sort of hierarchy determining the order of creation of the elements, since they cannot be made simultaneously. Certain elements will be given priority in the production process and will influence and determine the production of the elements made after them, and some will be shaping the production of the film more than others. Relative to one another, an element can either be a *controlling* element or a *controlled* element.

The *controlling* element, created first, is more difficult to change, control, reshoot, or duplicate, and is usually more complex than the elements it controls, while the *controlled* element is created and designed to fit existing elements, and is usually made by processes which provide a higher degree of control and flexibility in the construction of the element. For example, in *Who Framed Roger Rabbit*, the live action was created first, and the animation added afterward; the animators had to adapt the actions and timing of the animated characters in order to complete eyeline matches or cover devices hidden behind the characters. After the shot or sequence is designed, all the elements can be prioritized; the easier an element is to control, change, produce or reproduce, the lower on the list of priorities it becomes. Less flexible elements gain priority, and the shot is built around them.

Within classical narrative film practice there are many working methods that privilege certain elements over others, often for technical or aesthetic reasons (for instance, lighting is often subordinate to camera placement, which in turn is subordinate to narrative motivation; and in most films, the soundtrack is added after picture lock). Productions with meticulously planned effects shots are less likely to allow improvisation on the set, composers often must take sound effects into account when scoring a film, and an editor's choice of takes for a particular shot may be influenced by effects demands.

Other film practices can have different hierarchies of elements; for instance, in music video, image is structured by sound (as is true in some animated films and TV commercials). In documentary or televised news, sound can structure the image, old stock footage can structure new footage, and live reporting can structure the program; and avant-garde films frequently derive their structure from a variety of sources. No matter how a film is made, there will be some sort of hierarchy determining the order in which the elements are made and combined together, and also which elements will be given priority if an aesthetic or technical conflict arises. Hierarchies may vary, but they are always present, affecting the making of every film.

Compositing demands that the shot be broken down even further, using the *element* as the basic unit of study, because several difficulties can arise when the "shot" is used as a basic unit. The *element* is already the basic unit of production, each being planned and created separately whether it is the entire image, a background, an animated figure, a matte or travelling matte, and so on. Whereas shots are, by definition, not simultaneous, the use of the element as a basic unit could provide an opportunity to examine simultaneous and spatial relationships between aggregate parts. The fact that multiple elements can appear and overlap makes the structuring of the image track more like that of the soundtrack, and brings theoretical models of the sound and the image track closer together.

The difficulties arising from the use of the "shot" as a basic unit can be grouped into several categories: (i) single-frame series that suspend continuity but act together to create a Gestalt, (ii) editing within the shot or within the frame, where cuts can be spatial instead of temporal, (iii) "masked" transitions which hide or erase the boundaries between shots, and (iv) "single-shot" films, which appear to be one long take or a single shot (although I will use the term *film* consistently here, most of what is said here applies to video as well). Each of these categories also represents a level of structure affected; the first deals with frames, the second with shots, the third with transitions between shots, and the fourth with entire films; often, a larger structure can contain a smaller one within it, and may even use it in the creation of its overall effect (for example, (iv) could contain a series of masked transitions, or (i) could be occurring in one third of the screen in a split-screen triptych). At different levels of structure, these effects destroy the shot's boundaries internally or externally; (i) and (ii) fragment the shot into smaller units, while (iii) and (iv) blur shots together into larger units.

In a single-frame sequence, consecutive frames may have little or nothing in common and produce a kind of flicker effect, momentarily suspending continuity. Yet viewed together, the sequence of frames creates a temporal Gestalt which unifies the sequence. These series are found most often in animation, the mode of production most concerned with individual frames. One example occurs in *Somethin's Cookin'* (*Who Framed Roger Rabbit*'s opening cartoon), when Roger gets electrocuted with his fingers stuck in wall outlets. The single-frame sequence occurs over 108 frames, and by stepping through these frames on a laserdisc player, the individual images can be seen. Many of them use different backgrounds, three are completely white, and over 40% of the frames do not even contain Roger at all. These flash frames, working together and with the soundtrack, create the impression of a continuous shot, and no one notices Roger's disappearance.

Films unmotivated by seamless realism or transparency, like Peter Kubelka's *Arnulf Rainer* (1960), films by Norman McLaren, or any of Paul Sharits' flicker films, also function in a similar way, using a temporal Gestalt to unite disparate pieces in an overall effect. Some of these films contain sets of alternating frames using persistence of vision to create what almost appear to be superimpositions, differing only in their flickering quality (the effect is the same as that of the Thaumatrope).

Superimpositions, split-screens, matte work, dissolves, and combinations or multiples of these are examples of editing within the shot (or frame); one could call this spatial editing, as opposed to temporal editing. Of the four categories (i) through (iv), editing within the shot or frame is the most widespread and examples of it are quite common. The formal and narrative possibilities of this kind of editing are still far from being exhausted, and combinations of techniques can produce interesting effects; for example, a split screen could show different locations, and a composited object or character could cross over from one location to the other. With spatial editing, the articulation of space can change constantly from three-dimensional to two-dimensional and back, and other graphic concepts like figure and ground can be experimented with.

As mentioned earlier, editing within the shot is changing over to digital technology, with its greater flexibility and "cel-layer" approach. The bias towards digitization is also one that encourages manipulation, since digital provides the greatest degree of it possible, not only for manipulation within the shot, but for manipulation between shots. Once the image is broken up into individual elements which can appear, disappear, and move independently of each other, these elements can be used to link shots together by having one or more elements carry over the cut from one shot to another. Frequently, this technique is used to hide or erase the cut altogether, making space and time appear as continuous as it does within a single shot, and resulting in what could be called a "masked" transition.[22]

Whereas normally a straightforward cut separates two shots, masked transitions use elements to seamlessly join two shots together without any temporal ellipsis or change of viewing angle, and these elements do not obey the boundaries which are used in defining where one shot ends and another begins. *Who Framed Roger Rabbit* has a number of these transitions (which are especially easy to achieve through animation). The most noticeable example is the one which joins the cartoon world to the live-action one, with the opening of the refrigerator door; one shot is animated, where the refrigerator lands on Roger's head, and the other is live action combined with animation, where the door swings open and Roger is joined by actors in a live-action studio. The action, sound, and visuals are designed as completely continuous, and there is only a change from a white to an off-white screen where the transition is made between the animated and live-action diegeses. Similar techniques have been used in other live-action films, such as Hitchcock's *Rope* (1948) where dolly-ins to dark areas are used to mask transitions between shots, and in numerous music videos such as Peter Gabriel's *Sledgehammer*, where animated sequences are seamlessly joined through the careful control of individual frames.

Masked transitions can also be used to join two spaces together; one such example can be found towards the end of *Somethin's Cookin'*, in the scene where Roger fills with air and flies through the kitchen. Although the space he flies through appears consistent, it can be clearly seen (in a frame-by-frame examination) that two animated backgrounds have been joined together to appear continuous. Although the kitchen cabinets in the background of the scene are drawn so as to fit together, the line between the two backgrounds is clearly visible, and there is a difference in color saturation between the two. While relatively common in animation, this onscreen joining of shots also appears in live action, most noticeably in a number of HDTV productions, particularly the music videos directed by Zbigniew Rybczynski. In one for John Lennon's *Imagine*, different takes of a single room are edited together side by side so as to appear as a series of connecting areas. This effect is easy to achieve with digital compositing, but it could have been done optically as two push-offs designed to appear as adjacent areas.

Another way of joining two spaces or backgrounds, even incongruent ones, is to have one of the elements grow in size so as to fill the screen during which time the background is changed behind it. Afterward, the element shrinks back to a smaller size and the transition is complete and the appearance of spatial continuity is maintained. Two examples of this appear in the *Somethin's Cookin'* sequence; in one, the rolling pin rolls off the counter, growing larger and filling the frame, a change of background drawings is made, the rolling pin shrinks as it falls away to the floor, the action appearing continuous throughout. Another transition occurs where Roger is flying around the kitchen; he flies into closeup with his lips blubbering, and flies away, the background having changed during his closeup.[23] Masked transitions attempt to build long takes out of two or more shots, and this process could be continued (as it is in a few of the works mentioned earlier) to the point where the entire film appears to be one seamless piece; when this happens, the film enters the realm of single-shot films.

A single-shot film is one which does not contain cuts, in the conventional sense, appearing instead to be one long shot. Not surprisingly, most of these works are found in cinema's more marginalized areas; the avant-garde, music videos and HDTV videos, computer-generated films, experimental animation, and even some television commercials. Several different approaches can be used in avoiding the conventional segmentation of cuts; the simplest of these is to record all the action in one long take, during which the camera runs continuously. Another approach takes a number of shots or elements and edits them together to *appear as* one take, obscuring the borders between them. A third method avoids using a camera entirely; these "camera-less" films can be created on a computer and printed directly on the film, like the computer animation used in IMAX films; they can be created on video, like some of Ed Emshwiller's work; or made through direct manipulation by hand, like Stan Brakhage's *Mothlight* (1963), Len Lye's *Colour Box* (1935), and many films by Norman McLaren.[24]

These four categories demonstrate how compositing and related practices problematize conventional concepts of the *shot* and the *cut*, producing forms that require a breakdown into individual elements, hierarchically ordered and combined into the finished product. Although their origins may lie in more marginalized areas, the techniques outlined here are being used increasingly in network television, and with the proliferation of special effects, in mainstream commercial film.

The ease of computer compositing is allowing more interaction within the image, and the independence of the elements allows image compositing to take shape in real time according to the user's commands in computer programs, video games, and virtual reality, and video cameras with compositing capabilities are now available to home video consumers. Industrially, the transition to digital has advanced another step with systems like Eastman Kodak's Cineon digital film system,[25] and the advances in image manipulation technology they represent, particularly in the areas of resolution, speed, and storage capability, have marked the beginning of a new era in filmmaking and special effects cinematography.

Compositing will continue to call into question our relationship to the image, and how it is constructed by filmmakers, received by audiences, and conceptualized by theorists. Theories dealing with editing and the shot need to be reexamined to account for fragmentation taking place within shots, elisions between them, and the effects of new technologies on image creation and manipulation. These, in turn, mean that notions of space, time, unity, and audience involvement need to be reconsidered, from both production and reception standpoints, to account for the high degree of articulation occurring in composite imagery, and the widespread use of these techniques. This is especially necessary with the rise of digital compositing, which allows control of the image down to the tone, saturation, and hue of the individual pixels.

These techniques are by no means limited to entertainment and fiction films; they are in use in the production of mass media as well. For example, digital retouching of imagery occurs quite often in the magazine industry, even in those publications devoted to documentary efforts like *National Geographic*.[26] Of course, retouching is as old as the photograph itself; but digital media, as we have seen, are deliberately designed with such flexibility in mind. Applied to a world in which much communication occurs through digital media, this flexibility can become a means of compositing fiction and nonfiction until they are indistinguishably fused.

NOTES

1. Color filmstock itself is really a composite of as many as twelve different layers of emulsions, and, as stated in Eastman Kodak's pamphlet *The Moving Image*, "every color negative film has at least eight emulsion layers, and most of them have nine." (page 7).
2. The Acme-Dunn printer was the first commercially produced optical printer, manufactured for the armed forces during World War II and introduced to the professional film industry after the war. It was designed by Linwood Dunn and Cecil Love, and won a technical Academy Award in 1944. See Raymond P. Fielding, *The Technique of Special Effects Cinematography*, 4th edition, Boston, Massachusetts, and London, England: Focal Press, ©1985.
3. Many camera moves are not possible in production due to the fixed perspective required, and panning is only possible with a nodal point tripod, which can maintain perspective across an arc. In postproduction,

pans can be done under certain circumstances; ILM has a method in which an image shot on 65mm is panned across and rephotographed onto 35mm. See Smith, Thomas G., *Industrial Light and Magic: The Art of Special Effects*, New York: Ballantine Books, ©1986. For ways in which lighting is effected by compositing —and vice versa— see Cathy Stephens, "Lighting for Trouble-Free Compositing", *ON Production and Post-production*, October 1994, pages 30-35.

4. Pizzello, Chris, "Projecting Realism With Introvision", *American Cinematographer*, December 1993, page 66.

5. Ibid., page 67.

6. From *Dialog: Industrial Light and Magic*, an interview conducted November, 1990, and published by Eastman Kodak Company, ©1991, page 4.

7. Even when a viewer does not know both languages, differences can be obvious. When, for example, a character in a Kung Fu film shouts an angry 30-second long response, and the subtitle is merely "Damn you!", it is obvious that something is missing. Also, viewers may recognize cognates in the foreign tongue which do not appear in the translation.

8. Aerial-image printing can be done on an animation stand with a projector that projects an image from below and focuses it onto the bed of the animation stand, while the camera above the bed, focused on the same plane, records the images. Cels and mattes are then placed on the bed, blocking out areas of the projected image or compositing other artwork or images into the shot.

9. Fielding, Raymond, *The Technique of Special Effects Cinematography*, 4th edition, Boston, Massachusetts, and London, England: Focal Press, ©1985, page 134.

10. From an interview conducted on June 21, 1991, with Jane and Dale Baer of Baer Animation, in their studio in Burbank.

11. See the "Dreamquest Branches Out" section of the "Production Slate" column compiled by Marji Rhea in *American Cinematographer*, December 1993, page 16.

12. Examples of this are morphing sequences found in many films and television commercials; as one element morphs into another the boundary between them is completely erased. Quite often, live-action elements are altered by computer and smoothly recombined with their backgrounds, leaving no distinguishable boundaries between elements.

13. An effect appearing in a wide variety of films including *Sherlock Jr.* (1924), *Meshes of the Afternoon* (1943), *The Parent Trap* (1961), *Futureworld* (1976), *Dead Ringers* (1988), *Double Impact* (1991), *Star Trek VI: The Undiscovered Country* (1991), *Dave* (1993), *The Last Action Hero* (1993), and *Timecop* (1994), to name but a few. At least in these productions, the actor's double appears as another character; but some video, optical printer films and HDTV projects use multiples of the actor for purely formal reasons, lining up synchronized series of them Busby Berkeley style, exchanging any notion of 'characters' for that of 'graphics'.

14. From an interview conducted with Robert Zemeckis at the University of Southern California, on April 17, 1991. In *Who Framed Roger Rabbit*, the rules for Toons can even affect humans; Eddie Valiant discovers this during his visit to Toontown.

15. Most of the top-grossing films have been ones making heavy use of compositing; *Titanic* (1997), *Star Wars Episode I: The Phantom Menace* (1999), *Jurassic Park* (1993), *E.T.* (1982), *Star Wars* (1977), *The Empire Strikes Back* (1980), *Return of the Jedi* (1983), *Raiders of the Lost Ark* (1981), *Indiana Jones and the Temple of Doom* (1984), *Ghost* (1990), *Forrest Gump* (1994), etc.

16. During the shooting of pickups for *The Addams Family* (1991) Christopher Lloyd was reportedly receiving $150,000 a day; to save time travelling to locations, he was shot as a bluescreen element and composited into the shots later.

17. From an interview conducted with Robert Zemeckis, at the University of Southern California, on April 17, 1991.

18. The entire Toontown sequence is a good example of this; Bob Hoskins was filmed alone on Industrial Light and Magic's bluescreen soundstages, and later added to the animated backgrounds. However, digital compositing, which can control individual pixels, allows actors to stay in the shot, and simply erases or manipulates parts of their image (provided that a plate of the background behind them has been shot and can be substituted in if needed).

19. Optical effects occasionally were used to hide mistakes; Richard Williams, one of the head animators who worked on the film, explained "Ken Ralston [of ILM]... He's the unsung hero... he did the final marriages [composites]. So, if we had a foot bubbling around on the floor, that somebody couldn't draw, Ken would just put a shadow on it." (Quote taken from an interview with Richard Williams conducted by Kathryn Kramer in London on August 1, 1991.)

20. *Terminator 2* is also senselessly violent, yet pretends to be against violence, it attacks a large technological corporation, and yet is among the most expensive movies ever made, with a budget around $110 million.

21. Force feedback suits can simulate touch and resistance of objects that the user interacts with, and during these interactions the computer can display the appropriate limbs of the user interacting, allowing the user to see his limb as a computer graphic which is controlled (or appears to be controlled) by the user.

22. The term "masked transition" was developed collaboratively, by myself and others, in CNTV691, a class taught by Marsha Kinder and held at the University of Southern California in Spring of 1991.

23. Several background changes occur while Roger flies through the kitchen; besides the one in frame 05882, there is one at 05925, where Roger's face fills the frame allowing the background to be changed, this happens again at 06013 to 06021. In frames 05938 to 05941 and again in 06035, there is a change in color saturation and a slight shadow where one cel layer overlaps another in the background, showing the seam worked into

the checkerboard design of the floor. (Frame numbers refer to the first release of the Touchstone Home Video CAV letterboxed laserdisc of *Who Framed Roger Rabbit.*)
24. The elements in camera-less films are usually not photographic elements. If one defines an element as one of the aggregate parts of which a film is made, then the elements in a film using direct manipulation might include drawn lines, dots, scratches, punch holes, and so on (even moth wings, in the case of Stan Brakhage's film *Mothlight*). In a computer-generated film, even the notion of the element becomes difficult to apply, since they are mathematical models and not physical materials.
25. According to a press kit distributed by Kodak, the following are the image dimensions for the formats supported by the scanner and recorder, each of which can process a standard 35mm frame every three seconds; for Super 35, with an aspect ratio of 1.32:1, 3112 by 4096 pixels; for Academy format, with an aspect ratio of 1.33:1, 2664 by 3656 pixels across; for Cinemascope, with an aspect ratio of 2.36:1, 3112 by 3656 pixels across; and for VistaVision, with an aspect ratio of 1.50:1, 4096 by 6144 pixels. The press kit also notes on page 3;

> ...the system is designed to accommodate 10 bits of resolution per color record, expressed in printing density. This is the familiar logarithmic metric of the film industry. Additionally, it provides a compact notation for the extremely wide dynamic range of film. The system will also have the capability of scanning, storing, and processing 8 bits per color.

26. See Ritchin, Fred, "In Our Own Image: The Coming Revolution in Photography", New York: Aperture, ©1990. Throughout the book he gives examples of digitally retouched photographs appearing in magazines, and he often recounts the methods used as well.

III.

Communication Media

6.
Machine Mediation of Social Interaction

... That the Atlantic and Pacific telegraph would be the source of infinite satisfaction to thousands of our hardy western pioneers who, through it, would be enabled to communicate with their wives and children, friends and relatives at home, need scarcely be mentioned. Many a heart would be gladdened, many an expense saved, and many a comfort added to scanty means, by early tidings of the emigrant's new favorable location and success. In whatever light the subject may be considered, whether in reference to the interests of the government, the prosperity of our merchants and navigators, or the happiness and comfort of the citizens at large, the enterprise is eminently calculated to promote the power, wealth, and general prosperity of the country.

—Ephraim Farley, on behalf of the House of Representatives Committee on Territories, in a report to the 33rd Congress, December, 1854.

...we are the future—more comfortable with things than people... That's definitely the direction we are heading as a society.

—Taku Hachiro, *otaku* and author of *Otaku Heaven*

The electrical telegraph was the first means of communication which was completely machine-mediated. Earlier means of long-distance communication could be heard directly (such as drums), or were visible to the naked eye (like the optical telegraph and smoke signal). Messages passed from one person to another were either direct, or in the case of letters by courier or a chain of optical telegraph stations, indirect but still delivered by another person. Messages sent over the telegraph could travel any distance without an intermediary and were not directly readable by the unaided human senses; machines at either end were needed to encode or decode them into something human beings could hear or see. For the first time, messages were completely invisible and inaudible while en route.

The shift from the postal letter to the telegraphic signal is analogous to the shift from analog to digital form; the message transmitted becomes a signal abstracted and detached from its physical context or container. A telegraphic message, decoded into a series of words (or, in the case of messages sent on the optical telegraph, indicating a choice from a range of possibilities), arrives in the barest form possible, whereas a letter, as a physical object, often contains additional information (objects sent with the letter, even the stationary itself) and may contain some form of inflection, when hand-written by the sender or when the paper is scented with perfume (one could hardly imagine sending a heartfelt *billet-doux* by telegraph). Moreover, the letter's status as a physical object enables it to convey other information about its history, such as stains, tears (or *tears*), dirt and wrinkles, that it may acquire during transit —damage which, if sufficiently extensive, can obscure the message being sent, just as noise can obscure an electronic signal. The letter was a private communication; unless the sender owned a telegraph and knew Morse code, messages sent by telegraph would be read and keyed in by the telegraph operator. The telegraph was also used mainly for long-distance communication, and was not something that most people used on a daily basis.

The telephone was more private and required less encoding on the part of the sender; one could simply speak and the machines (and human operators) would do the rest. As the first widely available means of machine-mediated communication, the telephone had a tremendous impact on social interaction. It changed the nature of communication, setting the foundation (economically as well as socially) for the telecommunications industry. New forms of etiquette developed for telephone usage, the technology brought different classes and cultures into contact, and the connections it made possible between individuals allowed new social patterns to emerge. Technologies such as the fax, electronic mail, networks, and video transmission are responsible for similar effects on contemporary society, and are also intimately involved with the interaction they are designed to mediate.

While the telephone represents the communication revolution brought about by analog technology, the computer best represents the revolution brought about by digital technology. The translating of messages into ones and zeroes makes them machine-readable, allowing machine control to be used for everything from database search mechanisms to command messages directed to the computer itself. Once again, the nature of communication and interaction have changed.

Through technology, communication and social interaction have given way to machine mediation. Group interaction becomes fragmented into a set connections between individuals (e.g., in electronic mail on networks where these links are forged); while this mobilizes the members of the group, it also allows them to disperse. Although technology allows connections that might not have otherwise been possible, connections between individuals grow tenuous, with layers of mediation through which communication is made to pass, and mediated global connections can displace or replace direct local ones. As the numbers of layers grow and machines take on greater importance as mediators, communication itself begins to change, along with the notion of what it means to interact. Technology has also given rise to the notion of "interactivity", which must be examined, as well as the process by which communications *through* machines can become communication *with* machines, as computer interfacing grows increasingly anthropomorphized.

The computer's capacity for mediation is greater than that of all the other machines preceding it, and thus it can be used to restrict communication as well as to foster it. How that mediation is used, and what form of it is chosen, tells us something about the person or institution using it; whether they answer their own e-mail, or automate it, or whether they even have an account at all. Some people refuse to get answering machines, and some do not even like phones, while others will have pagers or beepers or carry cellular phones with them. And there are still letters and postcards. With a wide variety of media to choose from, the process of sending a mediated message becomes a kind of communication in itself.

Group Interaction vs. Groups of Interactions

Like the telephone, television's appearance in the home also changed patterns of family life. In Barry Levinson's film *Avalon* (1990), there is a scene which brilliantly and succinctly sums up the situation. A family is eating supper while watching TV. Each person sits in a chair, silently eating a TV dinner on a tray, attention fixed on the television they all sit facing. On TV, a family sits around a table, facing each other, chatting happily as they eat dinner together.

Levinson's ironic image could be extended today to a lab full of computer users, each sitting quietly at their own computer, sending e-mail or taking part in an on-line Multi-User Dimension (MUD), where multiple participants communicate together simultaneously using onscreen text. Machine-mediated communication is so common that it is, to some people, unthinkable *not* to have a phone, or today even an answering machine —or for some, pagers, cellular phones, and electronic mail. Even if one doesn't want to use such systems, there are often enough other people who do that one is forced to use the system to reach them. With all the communications systems available, it is quite common to focus on the global at the expense of the local; many people have regular contact with friends in other states or countries and yet do not even know their next door neighbors. Quite often, the mediated global is preferred over the unmediated local; the mediated nature of the communication is overlooked in favor of the McLuhanesque "Global Village" it supposedly brings about. But while contact may be global, it is certainly not like that of a village. As philosopher Jacques Ellul writes,

> What characterized the traditional village was that everyone knew everyone else. It was global person-to-person knowledge. Information was transmitted by an individual and was accepted as more or less serious and important, depending on the way that individual was perceived. The story of the shepherd boy who pretended to be attacked by a wolf is significant in that respect. There is no global village today because I do not personally know the personality, the nature, and the ideology of the individual who transmits the information to me... The infinite communication networks do not bring me closer to anything or anyone. Although the telephone allows such personal contact, the human relationship is greatly altered when information is transmitted through that technology.
>
> Information transmitted by humans is not the same as that transmitted by bees or ants, although a risk of the worldwide communication networks is that it is to be reduced to that...
>
> The same sentence uttered by the president of France and by me is not really the same sentence. We disassociate words from the individual who says them in modern communication systems. It is impossible to do otherwise. When I listen to a TV news report, I don't know whether or not the news organization is connected to the political power being described, nor can I establish whether if the advertising sponsors influenced the way the facts are presented. I have an immediate source, but does that person act independently or not?[1]

With changes in the means of communication, group interaction is frequently replaced by a group of interactions. Group interaction in which members are physically present is more than merely a simultaneous set of interactions between individuals; it is synergistic, the whole more than the sum of its parts. Members of a group interaction are often aware of what everyone is doing at once, not just the person they are speaking or listening to at any given moment. The environment of a group interaction is context-rich and shapes every individual interaction that is a part of it; the interactions involved collectively shape each other and determine the overall direction of the group's activities. Likewise, the individuals involved must often give their undivided attention to what is going on, perhaps reading subtexts such as body language, posture, and facial expressions as well. (Beyond that, one could even add such things as pheromones, scents, tones, manner of dress, height, weight, and a multitude of other subtleties that may register only unconsciously, but contribute nonetheless to an overall feeling or impression of a person or the atmosphere of the proceedings.) During machine-mediated interaction, participants are usually alone, reacting to the representations of the other participants as they appear on screen or speaker.[2]

Machine-mediated interaction is undeniably less socially and informationally rich than a face-to-face, in-person encounter, and depersonification of the person on the other end of the line is always a danger. Narrow communications channels and machine representations have even been shown to diminish emotional expression.[3] But this need not always be seen as a negative trait; mediated communication may be more useful in situations where logical or intellectual response is desired over an emotional one. The context and the task for which machine-mediated communication is used play a large role in determining where and when its use is appropriate. Studies comparing computer-mediated communication to face-to-face encounters have yielded a wide variety of results, showing machine-mediated communication to be a very complex issue determined by many variables, such as familiarity with the technology used, the task at hand, and the individuals using the technology. Experiments have occasionally produced conflicting results; in some brainstorming was found to be better when done through computer mediation, with larger groups outperforming smaller ones, while other experiments found face-to-face groups to be more productive.[4]

Communications researchers have reached a wide range of conclusions, which are continually being appended. Computer-mediated discussions may have more delays, more outspokenness and "flaming" (malicious comments), and reach more extreme, unconventional and risky decisions, but they also allow the redistribution of work time and the participation of peripheral workers. Several studies point out how computer-mediation has an equalizing effect on its participants, allowing them to be less inhibited and more critical, and how it reduces social pressures to conform to majority judgements. Other studies revealed interesting effects of computer mediation: face-to-face groups are more likely to come to a consensus and opinion change (or, perhaps, conformity), but on the other hand, computer-mediated discussion was found to have less argumentation, and decisions made in computer-mediated groups were found to have shifted further away from the individual members' initial choices than the group decisions following face-to-face discussions. Intergroup interactions were found to be dramatically more competitive than individual interactions.[5]

Whatever the case, machine-mediated communication tones down the emotional content of an exchange (often focusing more on the idea or transaction at hand), allows monitoring to be done by an outside party, and makes communication possible wherever the participants (and machines) might be. These factors, along with the reduction in overhead that less office space means, have attracted corporations to the idea of the home office and the mobile office.

The mobile office is made possible through the combination of such technologies as the cellular telephone, laptop computer, facsimile (or fax) machine, and other miniaturized computer hardware like scanners and printers. Mobile computing units and personal digital assistants often attempt to integrate several of these technologies together. BellSouth Cellular's "Simon", for example, had office software including an address book, file system, electronic mail, alarm clock, notepad, calendar, sketch pad, cellular paging system and multiple calculators, and a built-in fax modem. These units are linked into wireless wide area networks (WAN), metropolitan area networks (MAN), or local area networks (LAN). WANs cover the largest areas, and are run by satellite services and radio paging services; private radio networks, like RAM Mobile Data and ARDIS run the wireless MANs; and LANs cover small areas like single buildings or office campuses. Through these networks, workers can communicate with each other and the corporate home base. The market for mobile computing hardware is in the tens of billions of dollars, and there are millions of people joining the ranks of mobile office employees every year. In November of 1993, Richard Dalton reported that a study using Bureau of Labor Statistics estimated that by the year 2000, 43% of the United States work force will be nontraditional workers, such as contract or temporary workers.[6] Nor are such changes limited to the West; developing countries are leapfrogging into state-of-the-art technology. Thailand, for example experienced rapid economic growth; as early as 1993, there were 1.5 million people on the waiting list for phone lines, and in early 1994, cellular phones made up 20% of the phone service there.[7]

Higher productivity can be a benefit of telecommuting, and some employees may like the freedom and greater responsibility that delegation of authority brings, but work environments can seem more temporary and less friendly as well. Downsizing of staff means more work for those who remain employed, and greater deadline pressures and longer hours can lead to greater workaholism.[8] As Robert Moskowitz pointed out, the mobilization of work means that employers will look for the cheapest workers, no matter where they are, and that "High-living yuppies in Aspen and Big Sur will be underbid by Indonesian, Indian, and South Americans with competitive skills." He adds,

> Speaking more psychologically and metaphorically, many critics of the information superhighway see it as a step towards virtual reality—which is another way of describing a step away from actual reality. They foresee a world in which more and more people are more and more isolated in their air-conditioned telematic workstations.

> They are isolated not only from flesh-and-blood contact with other people, but from contact with the planet itself. There is the potential not only for rampant disease as we lose our resistance to the planet's germs, but for individual and mass insanity as we lose contact with the boundaries of day-to-day life.[9]

Low morale among workers can also occur with the separation from office and co-workers, and, because work occurs outside the office, it can all too easily take up after hours time and invade the home, making the separation of work life and home life impossible, while saving the company the cost of office space and other amenities. The home office of today is a far cry from the shop owner who lives above the store. Unlike the shop owner, many mobile employees are not in business for themselves, and their working and living spaces overlap —often with business taking precedence.

Ironically, even bosses who do not telecommute may be bothered at all hours by telecommuting employees who have only a vague sense of what is expected of them. As Sue Shellenbarger notes in *The Wall Street Journal*;

> Another concern about virtual offices, managers say, is that some aspects of mobile employees' lifestyles are counterproductive. Karen D. Walker, Compaq's vice president, operating services, worries about getting her staff to stop sending faxes in the middle of the night. "People are now thinking and working on the job 12 to 18 hours a day," she says, adding that she tries to set an example by limiting her own hours.
>
> When Ms. Walker travels on business and finds employees' voice mail on the phone in her hotel room at 11 p.m., she explains, "I send them an e-mail that says 'Shouldn't you be doing something else at this hour?'" She has also had to ask two employees in the past three months to cut their hours. "Your mind," she told them, "has to have some downtime."[10]

While information technology has allowed the rise of the mobile employee, some forms of it have even begun to replace traditionally mobile employees. Encyclopaedia Britannica Inc., for example was badly hurt financially by the trend towards CD-ROMs for information storage and retrieval. This happened because the company listened to the demand of its powerful door-to-door sales force, who made commissions on the sets of books that they would not have made selling CD-ROMs; while multimedia CD-ROM encyclopedias sell for as low as $99 and are sometimes bundled with new computers, a set of Brittanica books costs $1,500 or more (they also require four and a half feet of shelf space and weigh 118 pounds, whereas Brittanica's 44 million words could fit on two CD-ROM discs).[11] Door-to-door sales have given way to telemarketing, which itself may be replaced by automated systems or interactive advertising. Instead of reaching a human operator, calls to businesses now often reach a recorded menu of touchtone choices that route them to voice mail to be retrieved later. Such recordings have made live phone connections seem much more "real" by comparison, even though "live" phone connections are themselves only electronic reproductions of the human voice. As this kind of technology grows so inexpensive that humans earning wages can't compete with it, companies will use face-to-face contact only where it benefits them; the telemarketer calls you hoping you won't hang up, but when you call in you get a recording and are asked to leave a message.

Layers of Mediation Between Individuals

Just like corporations, many individuals are looking for ways of avoiding unwanted contact; there are "caller ID" systems that let you know the phone number of the incoming caller before you pick up the phone. "Call Screen" allows you to screen out calls from up to ten different phone numbers, and "Priority Ringing" will let your phone "ring distinctively" for up to ten different numbers, so you can recognize who is calling before you pick up the phone. Even if a person does manage to get through, answering machines can be used to screen calls, and "call waiting" can even interrupt desired calls.

Overall, more and more social interaction is being mediated by machine. Of course, much of this interaction, such as long distance calling, would not have existed at all without technology, and without it many electronic mail correspondents would never have met. Still, no matter how one looks at it, the time taken to communication by machine *could* have been used for face-to-face communication —although not with the same people or results. It is quite a common phenomenon today for people to have regular contact with friends in other states or foreign countries, and yet not know their next-door neighbors. As distance communication allows distributed communities to form, local geographically-based communities may weaken; "living in the same area" does not seem to constitute "having something in common" as much as it once did. With the rise of distributed communities, individuals are in one sense more isolated, and more dependent on machines for their social interaction. On the other hand, the ease of machine-mediated communication can make people too accessible; pagers, beepers, cellular phones, and so on, can become an interruption and an annoyance. One television show host was even interrupted, while on the air, by a call from his mother on his cellular phone.[12]

As an article in the *LAN Times* notes, due to the use of these technologies concentration suffers, and important meetings get interrupted; and when the video phone arrives, people will have to look excited as well as sound excited.[13] Of course, pagers and beepers can be turned off, but people may fear missing a valuable call or important information. Priorities have become reversed; when talking with someone face-to-face and the phone rings, most people will make the person physically present wait while they answer the phone, and while talking on the phone, "call waiting" can allow incoming calls to interrupt the caller who called first!

Today, an increasing number of "conversations" taking place over the telephone lines are the text and picture files of electronic mail. Electronic mail, or e-mail, began in 1969 with the ARPAnet, which was developed in Boston for the Department of Defense's Advanced Research Project Agency. The ARPAnet evolved and grew into what is now the Internet, which at its 25th anniversary was estimated to have more than 30,000 subnetworks, 3.2 million host computers, and as many as 30 to 50 million users.[14] Although the popularity of computer networks did not take off until the widespread appearance of home computers in the late 1970s, their effects were already foreseen by some. In 1974, Hans Magnus Enzensberger wrote of how written culture superseded oral culture, and hinted at what might displace written culture;

> Electronics are noticeably taking over writing: teleprinters, reading machines, high-speed transmissions, automatic photographic and electronic composition, automatic writing devices, typesetters, electrostatic processes, ampex libraries, cassette encyclopedias, photocopiers and magnetic copiers, speedprinters. . . . Today, writing has in many cases already become a secondary technique, a means of transcribing orally recorded speech: tape-recorded proceedings, attempts at speech-pattern recognition, and the conversion of speech into writing.[15]

Enzensberger sounds as if he is describing electronic mail, and the systems and networks which have become the "newspapers" of digital culture. E-mail combines written and oral culture; it is used more for transmission than for long-term storage, the majority of it appearing and disappearing strictly in digital form without ever being committed to paper. While its messages are written on a keyboard and appear onscreen as text, its ephemeral, transient quality makes it more like oral communication, and it is often composed and discarded in the same offhand improvisational manner.

E-mail has proven to be a difficult medium to research and theorize, because it has such a wide range of uses, and many of its long-term effects remain unknown. As Jacques Leslie writes,

> Yet for all its alleged virtues, e-mail is a surprisingly quirky medium, so packed with idiosyncrasies that, despite a decade of effort, academic researchers have yet to devise a theoretical structure for understanding its impact on organizations. E-mail is written, yet its language typically embodies a shift toward oral speech patterns. It is the most ephemeral of written mediums, lacking the material form of books or letters and capable of being erased in a keystroke, yet it can be archived and retrieved with unprecedented ease. It has occasioned an astonishing effusion of warm-hearted social interaction outside work settings, yet at work it is the medium of choice for employees who don't like each other and wish to minimize their interactions. It is the glue that holds many virtual communities together, but some corporate managers report that they must supplement e-mail messages to subordinates with occasional phone calls to maintain personal relationships.
>
> And e-mail can just as easily promote workplace centralization as decentralization, for while workers may find that e-mail gives them unaccustomed access to resources throughout their company, their superiors enjoy enhanced means of tracking their performance.[16]

While the effects of machine-mediated communication such as e-mail (or even the telephone, for that matter) can be shown to have both negative and positive uses and effects, it is inevitable that such communication will have a lower bandwidth of information than face-to-face contact. This highly controlled and filtered form of communication may help some people to communicate, but it may also allow others to hide behind a computer and forgo face-to-face encounters. Whereas the telephone lacks facial expressions, e-mail lacks the inflection of speech, making its messages more ambiguous; even an answering machine message will contain inflection, tone, and response timing, indicating emotion, hesitation, or even health (if the person has a cold or sounds depressed). To remedy e-mail's emotional barrenness, the "smiley" has come into use in an attempt to provide the missing inflection. A smiley uses various punctuation symbols to create a sort of "face" turned sideways at the end of a message, to indicate emotional content. There might be a happy face :-) or a sad one :- (or any number of variations; David W. Sanderson's book *Smileys* lists hundreds of them.[17]

Not only does the limited bandwidth of machine-mediated communication act as a filter, giving a user greater control of what goes into a message, but it limits the sender's knowledge of any reaction to the message. Because the communications are machine-mediated, additional machines can be added to the chain of them between users, providing further filtering and allowing users to employ automation to avoid contact.

Automated response systems, whether on the phone or computer, are already in widespread use by industry, government, and universities, and are now commercially available for home use. Like the services available from the phone company mentioned above, these "digital filters" can not only limit (or eliminate) interaction between individuals, but between individuals and the outside world. Some, like *Archie* and *Rosebud*, can search for files or monitor streams of information, while others known as "bozo filters" can automatically block messages from people you don't want to hear from.[18] And the capabilities of these filters will increase; more than simply filters that act passively, these digital agents can be programmed to search data banks, files, and postings, or even to interact with other users. Suggesting one use, Nicholas Negroponte, founder and director of MIT's Media Lab suggests, writes in a June 1994 column;

> What I really need is intelligence in the network and in my receiver to filter and extract relevant information from a body of information that is orders of magnitude larger than anything I can digest. To achieve this we use a technique known as "interface agents." Imagine a future where your interface agent can read every newspaper and catch every broadcast on the planet, and then, from this, construct a personalized summary.[19]

Further along in the column, he chooses the selection of a movie as an example of what the agents might do;

> I am fond of asking people how they select a theatrical, box-office movie. Some pretend they read reviews. I hasten to interject my own solution—which is to ask my sister-in-law—and people quickly admit that they have an equivalent. What we want to build into these systems is a sister-in-law, an interface agent which is both an expert on movies and an expert on you.[20]

But how can an interface agent become "an expert on movies"? The agent will either rely on the opinions of people, or it will rate movies according to preprogrammed criteria which are probably shallower than those of a human reviewer. Nor does such an agent leave room for growth or fickleness; even if the agents are made to learn about a person's choices over time, there is no way for them to account for sudden shifts in tastes and interests, or include serendipitous discoveries of things outside of one's usual interests. And even if one programs a newspaper-searching agent on a daily basis, it could leave out new people, places, products, and stories appear which are unforeseen or previously unknown. And how will digital agents determine relevance? There is also the possibility that advertisers or competing news sources will devise ways of getting their material picked up by more digital agents, by examining the way they function —for example, adding keywords to which agents are likely to respond. At any rate, even if they *did* work as described, newer choices made by agents would always reflect past ones; the inherent assumption being that a person does not change much or does so along a predictable trajectory. A person could even more easily lose touch with or ignore things they didn't like, potentially becoming insulated and isolated.

Whatever the motive for using the limited bandwidth of machine-mediated communication, it always remains an abstracted form of human contact, and less rich than face-to-face contact. It undeniably has many advantages; communication is possible over long distances, people can keep in touch who might not be able to do so otherwise, and the controlled nature of the contact is often welcomed by the otherwise shy person. The limited information e-mail provides about its users means that people cannot be judged by their looks, but it also means people can misrepresent their race, gender, age and so on, sometimes with malicious intent. Racist and sexist messages, and even death threats are occasionally posted, some with the potential to be read by millions of users. Flaming is a common phenomenon, in which vitriolic exchanges are sent to intentionally start "flame wars".[21] Communication through the computer in which no other human being is actually seen creates psychological distance; the person on the other end is not thought of as such, nor are the millions of potential readers imagined. Because the interaction occurs over a computer terminal, it is easy to approach acts of internet violence with a video game mentality, without realizing how such acts affect others, and the very real consequences that can result —including jail. An Oklahoma court, for example, sentenced Oklahoma Information Exchange Sysop Tony Davis to ten years in prison with a $10,000 fine for "trafficking in obscene images"; and Matthew M. Thomas, a 19-year-old freshman at Stephen F. Austin State University sent a death threat by e-mail to President Clinton, and was soon faced with a maximum sentence of five years in prison, a $250,000 fine, and three years of supervised release.[22]

The deliberate avoidance and lack of practice of face-to-face human contact, due to shyness, anti-social sentiment, or isolation for whatever reason, can increase self-consciousness and decrease social awareness, resulting in social ineptness. In more extreme cases, it can distance a person psychologically and dull empathetic response, resulting in social dysfunctionality.[23] Furthermore, these effects are often coupled with the violent mentality found in many video games (which is increasingly more graphic as game imaging develops). As reported in *Wired*;

> Twinkie Defense Revisited: Did you hear about the kid who got off on charges of attempted murder by claiming he was driven insane by playing too much *Mortal Kombat*? We are not making this up. Paul Van Schaik, 16, of Fort Lauderdale, Florida, stabbed a classmate 17 times with steak knives (both hands!) in a demented quest to attain the next level in the gory videogame. Ruled legally insane and sent to a state institution, Van Schaik has provided a rallying cry for the anti-violence lobby.[24]

Although it is sometimes argued that certain individuals are already unstable, the isolation and distancing present in computer use and the violence present in the games often encourage their condition, as opposed to other more social activities that might help reduce it. In any case, while a person's behavior is overdetermined, making it difficult to prove or separate out and measure the effects of a particular influence, it cannot be denied that such effects exist, and that prolonged exposure has a desensitizing effect. And as the use of computers or video games grows into a time-consuming obsession, the number of *other* influences also decreases. Perhaps the greatest example of such effects can be found in the *otaku* of Japan;

> Dubbed the *otaku-zoku*, or *otaku* for short, these are Japan's socially inept but often brilliant technological shut-ins. Their name derives from the highly formal way of saying "you" in Japanese, much like calling a friend "sir." . . . The otaku are Tokyo's newest information-age product. These were the kids "educated" to memorize reams of context-less information in preparation for filling in bubbles on multiple-choice entrance exams.
>
> Now in their late teens and twenties, most are either cramming for college exams or stuck in cramming mode. They relax with sexy manga [Japanese comic books] or violent computer games. They shun society's complex web of social obligations and loyalties. The result: a burgeoning young generation of at least 100,000 hard-core otaku (estimates of up to 1 million have been bandied about in the Tokyo press) who are too uptight to talk to a telephone operator, but who can kick ass on the keyboard of a PC.[25]

The article goes on to provide a portrait of a typical *otaku* named Zero;

> Zero, 25, is a self-proclaimed otaku who flunked out of Keio University's math department because he didn't like being ordered around by teachers to whom he felt superior. "They couldn't deal with someone like me," he recalled. "Now I'm independent and don't need to deal with anyone like them."
>
> Zero's life now revolves around computer games. He only ventures out of his six-mat in Kawagoe to acquire new game-boards, the green, maze-like "minds" taken from commercial arcade games like *Galaga* or *Space Invaders*. At home, he plugs these circuit boards into a special adapter on his own console, analyzes and dissects them for bugs and flaws that allow one, for example, to glimpse a *space invader*'s afterimage as it scuttles across the screen or to change the color of a yellow *Ms. Pac-man* to purple.
>
> Zero often dresses in a plain white T-shirt and ill-fitting jeans rolled up about six inches. He doesn't look you in the eyes when he talks; he answers quietly with his face to the floor. His face possesses gentle features, but it is sickly pale.
>
> He makes his living as a software trouble-shooter, looking for problems in new software before it hits the market, earning ¥ 350,000 (about $2800) a month. He works in his murky home, where the windows are permanently covered with yellowing newspaper to block out the sunlight.
>
> "I've always liked playing games. As a boy, I preferred video games to other kids," Zero offered. "So I understand technology. I'm more comfortable with computers than human beings."[26]

From the article, it appears an *otaku* communicates mainly through electronic mail, which is clearly preferred over in-person visits. The technology (and technologically-oriented jobs provided by the computer industry) make their way of life possible, and the vast numbers of them even creating a kind of *otaku* culture —which, ironically, calls for personal isolation, and machine-mediation of any discussions that might occur. We might say, in a sense, that it is a form of "shared isolation"; a community that can *only* exist virtually, one centered not around its participants, but around their electronic representations.

The desire for power and mastery over one's surroundings may be a part of the reason why the predictable, controllable machine is preferred over other human beings, just as machine-mediated communication is sometimes preferred over face-to-face communication because of the control it affords. As the number of filters or chains of machinery linking individuals together grows, communication *through* the computer becomes communication *with* the computer, which serves as portal and mediator between the user and the world. At the same time, computer graphics, programming, and simulations can themselves occupy users for hours on end as easily as e-mail or other machine-mediated social interaction. Computers cannot "socialize" in the human sense, yet people often refer to them as "interactive"; it would seem necessary, then, to distinguish between interaction which is "social" and that which is not, and reexamine what is meant by "interactivity".

Interactivity

The phrase "interactive media" implies that traditional media are either passive, or active in a way that is not "interactive". Before the appearance of media, when everything was transmitted via human contact and word-of-mouth, interactivity was a given, and taken for granted; rhetoric surrounding "interactive" media often uses the term as a selling point. But what forms does this interactivity take?

The digital computer, with the stored program memory in which variables can be rewritten during the run of a program, has changed the meaning of the word "interactive". *The Oxford English Dictionary* gives two definitions of the word; the first, "1. Reciprocally active; acting on or influencing each other" is applicable to both scientific and social uses, and the references given for it go back to 1832. The second meaning, "2. Pertaining to or being a computer or other electronic device that allows a two-way flow of information between it and a user, responding immediately to the latter's input,"[27] is a newer definition; the earliest reference listed is from 1967. The second definition indicates that the nature of interaction has changed with the malleability of the computer memory. In other forms of interaction, such as in a social encounter between two people, or in the physical or chemical sense wherein two objects or forces affect each other, interaction produces long-lasting and irreversible results; although one person may forget another and materials can be replaced, the state prior to the interaction cannot be reproduced exactly. The computer, on the other hand, allows variables to be changed, but such changes are rarely permanent or irreversible; the zeroes and ones stored as ons and offs in memory are changed back and forth millions of times a second. Switch on a computer, load a CD-ROM and "interact" with it for hours, remove it and switch off the computer; despite the "interaction", the computer and the compact disc are as unchanged as a used newspaper (or less, since the paper will likely have more wear and tear). Shooting space invaders, choosing video clips and screens from menus, or turning a crank in *Myst*'s Stoneship Age may seem like interaction, but they are really closer in form to analogs for physical interaction; an image or representation of interaction. On the other hand, it is the human user who may be changed by the experience, possibly without realizing it. As the photograph detaches the image from its referent, "interactive media" which depict physical actions can detach them from their consequences, as in video games depicting shooting and killing. When actions are detached from consequences, so is the user's accountability for those actions.[28]

From a functional point of view, what constitutes "interactivity" in this newer, and popular, sense? Unlike traditional tools, "interactive" media are "active" in some way (one never hears of someone "interacting" with a shovel or hammer or eggbeater); they are programmed to respond to certain "commands" (a typed word or mouse click). The idea of a series of choices in which early choices affect future options is also important; thus we do not consider those appliances requiring only a single choice to be made (such as the light/dark setting on a pop-up toaster) as "interactive" devices, nor those in which multiple choices are made simultaneously (such as the settings on a washing machine), nor even something like a remote control, in which choices do not limit future options. Ironically, communication technologies that mediate social interaction such as the telephone, e-mail, or closed-circuit video, are typically not included in the realm of "interactive media" although they are certainly more interactive than the video game and CD-ROM since they allow one to interact with other people, not a computer program. However, the distinction between these types of interaction is less clear as one looks closely at the different types of interaction possible.

Looking at the various configurations of people and machines, we can discern seven types of interaction, all points along a smooth continuum, and notate them as follows (solid lines indicate primary connections, while dotted lines indicate secondary connections):[29]

(Two Person, Direct)

1. Person1 ←————————————→ Person2

2. Person1 ←——→Machine←——→ Person2
←- - - - - - - - - - - - - -→

(Two Person, Machine-mediated)

3. Person1 ←——→Machine←——→ Person2 (real-time)

4. Person1 ←——→Machine←——→ Person2 (delayed response)

(One Person and Machine)

5. Person1 ←-----→Machine (Algorithm ← - written by Person2)

6. Person1 ←-----→Machine (Algorithm ← - written by Person1)

(Machines Only)

7. Machine1 ←-----→Machine2

The first two cases involve people who are both physically present in the same space; in type 1., contact is unmediated person-to-person, face-to-face, without machinery of any kind, direct social contact. In type 2., both people are physically present, but not all contact is face-to-face; some of it is through the machine that is present. An example might be two people playing an arcade or home video game for two (or more) players; each player has his or her own onscreen representation, which interacts with those of the other players. Besides the mediated interaction onscreen, there is direct interaction going on as the players talk and turn to look at each other; the amount of direct interaction can vary from almost nothing to the ignoring of the machine altogether (at which point it would become like that of type 1). Most video gaming, however, requires more concentration on the screen than on the other person, and so it is not unusual to find that the greatest amount of interaction occurs through the machine.

Supposing that the second person is not just across the room, but in another room, or even far away? The next two types, 3. and 4., involve two-way communication between two people, but communication entirely mediated by a machine (or series of connected machines, which I will consider as all part of the same system; thus your telephone, my telephone, and all the lines and switchers connecting them can be considered as a single system —another way of looking at it would be to break down the chain into a series of interactions of type 7; between your phone and mine is a series of machine-to-machine interactions joining the two together). In type 3., the mediation allows communication to occur in real-time, at (nearly) the same rate as it would in person (the reduced bandwidth notwithstanding); examples of this might be the telephone, radio communicators, closed-circuit television, or the "chat" mode in electronic mail in which each person's text appears on each person's screen at nearly the same time. These forms of communication may appear to take place in real-time, but none of them are truly instantaneous. As distance increases, so does response time; it takes roughly one second for a radio signal from earth to reach an astronaut on the moon, for example.

As response time increases, due to distance or "time-shifting" technologies such as answering machines and electronic mail which can hold messages until they are retrieved, interaction grows closer to type 4., wherein the participants are neither present in the same space or at the same time. As the social interaction fades into a series of delayed responses, the slower rate means fewer interchanges for any given length of time, and usually means fewer of them altogether. There is more time to think out messages and responses, but the greater control available can hinder spontaneity; communication becomes asynchronous message retrieval. With message retrieval, there is a reliance on a machine not only to *transmit* messages, but also to *store* them; in this sense, interaction with a machine begins to overtake human social interaction in the course of the communicative interchange.

As the layers of machine mediation increase, so does the interaction between human and machine, until finally this is the *only* type of interaction occurring, as in types 5. and 6. In type 5., a person interacts with a machine, or rather, with an algorithm written and programmed into the machine by another person; all possible responses are preprogrammed ahead of time. The goal of artificial intelligence (AI) research is a machine which can pass the 'Turing test", a test proposed in the 1950s by Alan Turing, a British Mathematician and computer scientist. A machine passes the test if a user communicating with it at a remote terminal cannot tell whether or not there is a person on the other end. Although it may seem rather unlikely that a machine could pass the Turing test, it is really a question of who or how many people could be fooled, and for how long; at a 1991 competition at Boston's Computer Museum, after a three-hour trial, one computer program, *PC Therapist III* from Thinking Software Inc., still fooled five out of ten judges, and won the $1,500 prize.[30]

The bandwidth, response time, and context of the communicative exchanges are also important factors (for example, one can already play chess or checkers through a computer without being able to tell if the opponent is human or machine). As such, the algorithm can take on the role of the other person and interact in place of its author. The person using the software may wish to respond to its author (Rand and Robyn Miller, for example, received many responses from people who played their CD-ROM, *Myst*), but more often than not game players and users of computer software do not attempt to contact the programmers. The author of the algorithm can simulate a presence in several ways, most commonly as the "computer player" against which the human player plays (the earliest commercially-available one being the computer player in *Pong*). The computer-controlled player represents an algorithm that the human player can sometimes find a way to beat (and which can even be programmed to cheat!), and with neural nets and fuzzy logic, computer-controlled players are even able to learn and improve their play. Another way the author's thinking can be present in the algorithm is in the design of the game itself. In *Myst*, for example, one must learn to think like the Millers, who designed the puzzles that are encountered. This kind of thinking could then be extended to include the design and interface of the computer itself. In type 6., interaction is almost wholly between user and machine, with the algorithm being programmed by the same user; computer programming is essentially human-machine interaction. The last type of interaction, number 7., takes place between two machines —a data transfer, control commands, and so forth, which are "interactions" in much the same sense as the interactions that occur between stellar masses or billiard balls, reverting back to the first of the two definitions.

It is easy to see how a society relying on one type of interaction could gradually make the shift to the next, since they form a smooth continuum. Together, they bridge the gap between interaction with a human being and that with a machine. Machine mediation allows users to be spatially separated, the ability to store messages allows for temporal separation, and advances in algorithmic complexity make possible increasingly complex simulations of communications; and digital technology allows all three functions to coexist in one machine.

In order to further smooth over the gaps between communication *through* machines and *with* machines, interfaces are designed to attempt to bring interaction with a computer closer to that of human social interaction. Interaction with early computers amounted to plugging in cables and reconfiguring hardware; gradually as the computer became more multi-purpose, programming changes could be limited to software alone. At first, programming could only be done through highly abstract machine code, but higher-level languages simplified programming into a vocabulary of commands. Today the majority of interaction with computers is not through programming, but the use of commercial software, designed with user in mind.

Social metaphors abound in computer interfacing; computers are said to be "user-friendly"; the first thing that appears onscreen when a Macintosh is turned on is an icon of a Mac with a smiling face (or the "sad face" if there are hardware problems); and even the word "interface" itself means literally, *between faces*, or *from one face to another*. Some technologies imitate less technical means of communication, including voice recognition and machines that read handwriting from a touch-sensitive screen. The combination of an interface that attempts to reproduce human interactions and the capability of programming to do increasingly detailed simulations of them, is resulting in a strange personification of the computer. As John Morkes reports in his article "The Leprechaun Effect";

> Is there a link between what you think about your computer and what you think about leprechauns? IBM, Apple, Microsoft, and US West are each paying Stanford University US$25,000 for a one-year membership in the Social Responses to Communication Technology to find out.
>
> Previous studies by researchers at the Department of Communication have shown that people attribute human personalities to computers. Now the researchers want to know if tiny faces on the screen elicit the same psychological responses as elves and leprechauns.
>
> "Many cultures around the world assign magical properties to people who are small," says Stanford Professor Byron Reeves. "These small people grant wishes, they monitor behavior, and they keep people safe. But they can also punish, or be bad just for the hell of it."
>
> "We want to know, when you see a small face on a screen, do you respond to it as if it were magical? Is it perceived as powerful or capable?" asks Stanford Professor Clifford Nass.
>
> The answers could have implications for the design of video conferencing systems, personal digital assistants, large-screen projection TVs, and any communication technology that represents humanlike beings in any medium, the researchers say.[31]

Personification causing emotional response can occur even without human imagery; *SimCity2000* lets players design and manage an entire city, complete with thousands of citizens who have to be kept happy. The 1990s also saw the spread of "Tamagotchis" and other "digital pets", pocket toys which featured tiny on-screen animals that one could feed and take care of by pushing buttons; without the proper care, the animals "died". Even more involved is *Aquazone*, a "virtual aquarium" program made by a Japanese software company. Describing this electronic fish tank, Kim Eastham writes,

> How real? My fish will die if I feed them too much, or too little. Or don't clean the water. Water temperature, purity, light, medicine—all are mine to control. Optional disks let me grow aquatic plants to entertain my guppies, platys, and angelfish.
>
> The fish grow, mate, and have offspring over months or years if you choose to draw out the drama (time frames are variable). The fish, water, and environment are so lovingly crafted, an untrained eye would take an Aquazone for the real thing. . . . Make no mistake, the 30,000-

> plus users of Aquazone are seriously into fish. The software has a library function with illustrated data on fish, plants, and parasites, care, and breeding.
>
> Better read up—and make a printout for the neighbors—'cause those little sparks of light might be lying at the top of the screen if you're called away on a business trip.
>
> "We had a couple who phoned us in tears when their fish died," said Takashi Mineyoshi, sales manager of Lits Compute in Tokyo. "People really become attached to this pet."[32]

One might ask how crying over the "loss" of electronically simulated fish differs from crying at the end of a sad movie. In the latter, it may be identification, an empathizing with the character; in the case of the fish, interactivity has created a feeling of responsibility; one does not take on a new identity or identify with a character, but one's actions do determine the outcome, and whether or not the "fish" continue to "live". Even though the fish have no more material existence than a space invader character, the context of "care and feeding" makes them seem dependent on the user, and a reflection of the user's sensitivity.

The effects of machine-mediated interaction, then, can range anywhere from the detachment of action and consequence, allowing 'flaming' and a desensitized video-game mentality towards violence and its depiction to occur, while on the other hand, it can provoke an emotional response and a sense of responsibility for personified constructs which are nothing more than the play of electrons inside a computer's circuitry. In any event, machine mediation made possible with digital technology has influenced not only the social fabric of society, but the emotional makeup of individuals, along with notions of what is considered "communication" or "interaction".

Although initially the technology served to split group interactions into a group of individual interactions, and blurred the line between machine-mediated interaction and interaction *with* a machine, the desire for humans to commune in larger groups remains. Current advances are working towards reintegrating these interactions into a visual whole, in the form of multiple user dimensions with graphic interfaces which attempt to emulate the group interactions that they replace. This, then, becomes the dream of a worldwide network, a dimension of pure communication, a dream most often referred to as cyberspace.

NOTES

1. Ellul, Jacques, "Preconceived Ideas About Mediated Information", in Everett M. Rogers and Francis Balle, editors, *The Media Revolution in*

America and in Western Europe, Norwood, New Jersey: Ablex Publishing Corporation, ©1985, page 103.

2. There are exceptions to this, for instance, two people in the same room who are talking to a third on a speaker phone (oddly enough, people in the room will often look at the speakerphone when the person speaks over it!). Machine-mediated interaction, however, is usually used because a participant is not, or cannot be, present, and so separation of participants is typically the case. On the other hand, there are some situations, as in two-player arcade or home video games, where two participants are physically present, but are concentrating onscreen where the player's representations appear, rather than on the other person (similar to Levinson's image).

3. Facial displays registering emotional response were found to be mediated by the extent to which individuals can fully interact in a situation. See Chovil, Nicole, "Social Determinants of Facial Displays", *Journal of Nonverbal Behavior*, Fall 1991, Volume 15, pages 141-154.

4. See Alan R. Dennis and Joseph S. Valacich, "Computer Brainstorms: More Heads Are Better Than One", *Journal of Applied Psychology*, August 1993, Volume 78, pages 531-537; Brent R. Gallupe, Alan R. Dennis, William H. Cooper, Joseph S. Valacich, et al., "Electronic Brainstorming and Group Size", *Academy of Management Journal*, June 1992, Volume 35, pages 350-369; Susan G. Straus and Joseph E. McGrath, "Does the Medium Matter? The interaction of task type and technology on group performance and member reactions", *Journal of Applied Psychology*, February 1994, Volume 79, pages 87-97; and Sara Kiesler and Lee Sproull, "Group Decision Making and Communication Technology", *Organizational Behavior & Human Decision Processes*, June 1992, Volume 52, pages 96-123.

5. Studies referred to here include: Kiesler, Sara and Lee Sproull, "Group Decision Making and Communication Technology", *Organizational Behavior & Human Decision Processes*, June 1992, Volume 52, pages 96-123; Michael Smilowitz, Chad D. Compton, and Lyle Flint, "The Effects of Computer-Mediated Communication on an Individual's Judgment: A study based on the effects of Asch's social influence experiment", *Computers in Human Behavior*, 1988, Volume 4, pages 311-321; Liora Bresler, "Student Perceptions of CMC: Roles and Experiences: II", *Journal of Mathematical Behavior*, December 1990, Volume 9, pages 291-307; Joseph S. Valacich, Alan R. Dennis, and J. F. Nunamaker, "Group Size and Anonymity Effects on Computer-Mediated Idea Generation", *Small Group Research*, February 1992, Volume 23, pages 49-73; Adrianson, Lillemor, and Erland Hjelmquist, "Group Processes in Face-to-Face and Computer-Mediated Communication", *Behaviour & Information Technology*, July/August 1991, Volume 10, pages 281-296; Timothy W. McGuire, Sara Kiesler, and Jane Siegel, "Group and Computer-Mediated Discussion Effects in Risk Decision Making", *Journal of Personality & Social Psychology*, May 1987, Volume 52, pages 917-930; Jane Siegel, Vitaly Dubrovsky, Sara Kiesler, and Timothy W. McGuire, "Group Processes in Computer-Mediated

Communication", *Organizational Behavior & Human Decision Processes*, April 1986, Volume 37, pages 157-187; and John Schopler, Chester A. Insko, Kenneth A. Graetz, Stephen M. Drigotas, et al., "The Generality of the Individual-Group Discontinuity Effect: Variations in positivity-negativity of outcomes, players' relative power, and magnitude of outcomes", *Personality & Social Psychology Bulletin*, December 1991, Volume 17, pages 612-624.

6. Dalton, Richard, "On the Road to a Home Office", *Information Week*, Number 453, November 29, 1993, page 63.

7. Owens, Cynthia, "The Developing Leap: Many emerging nations are moving directly into the new age, bypassing the wired stage", *The Wall Street Journal*, February 11, 1994, pages R15(W) and R15(E).

8. See Bartholomew, Doug, "The Longest Day", *Information Week*, Number 493, September 19, 1994, page 34(6); and "Are we being run over by the "virtual office" bandwagon?", *Telecommuting Review: The Gordon Report*, Volume 11, Number 7, July 1994, page 6(5); and Malone, Michael S., "Perpetual Motion Executives", *Forbes*, Volume 153, Number 8, April 11, 1994, page S93(4).

9. Moskowitz, Robert, "Telecommuting—A Growing Part of the Job Market", *MicroTimes*, August 8, 1994, page 43(4).

10. Shellenbarger, Sue, "Overwork, Low Morale, Vex the Mobile Office", *The Wall Street Journal*, August 17, 1994, pages B1(W) and B1(E).

11. Samuels, Gary, "CD-ROM's first big victim", *Forbes*, Volume 153, Number 5, February 28, 1994, page 42(3).

12. On the December 30, 1994, program of *Mutual Fund Investors* on CNBC, host Jimmy Rogers was interrupted mid-program by an incoming call from his mother on the cellular phone he kept in his pocket. Apparently, she was calling to inquire about his health and did not realize he was on the air.

13. See Harper, Eric, "In today's info age, how accessible is too accessible?", *LAN Times*, Volume 10, Number 11, June 14, 1993, page 100; and Canon, Maggie, "Unplugged: The merits of leaving technology behind while on vacation", *MacUser*, Volume 10, Number 10, October 1994, page 17.

14. Rendleman, John, "Business booming on the Internet", *CommunicationsWeek*, Number 520, August 29, 1994, page 1(2); and Leslie, Jacques, "Mail Bonding: E-mail is creating a new oral culture", *Wired*, March 1994, page 42. On the history of the Internet, see Birkhead, Evan, "Happy 25th to the ARPAnet", *INTERNETWORK*, Volume 5, Number 9, September 1994, page 48; and Brandin, David, "From ARPANET to Internet: out with the old, in with new", *LAN Times*, Volume 11, Number19, September 19, 1994, page 58(2).

15. Enzensberger, Hans Magnus, "Constituents of a Theory of the Media", in Hanhardt, John G., editor, *Video Culture: A Critical Investigation*, Layton, Utah: Peregrine Smith Books, ©1986, pages 119-120.

16. Leslie, Jacques, "Mail Bonding: E-mail is creating a new oral culture", *Wired*, March 1994, page 42.

17. Sanderson, David W., with text by Dale Dougherty, *Smileys*, O'Reilly Publishers, Sebastopol, California, ©1993.

18. Rheingold, Howard, *The Virtual Community: Homesteading on the Electronic Frontier*, New York: Harpercollins, and Reading, Massachusetts: Addison-Wesley Publishing Company, ©1993. On software agents, see pages 105-107. On "bozo filters", see pages 118-119.

19. Negroponte, Nicholas, "Less is More: Interface Agents as Digital Butlers", *Wired*, June 1994, page 142.

20. Ibid., page 142.

21. As an example, see Levander, Michelle, "A cyberspace gang fans flames on the Internet", *San Jose Mercury News*, May 15, 1994, on page 1E, about the "flame war" in which a group on alt.syntax.tactical sent graphic descriptions of cat-killing to a group of cat lovers. On racism on the net, see Levander, Michelle, "Slurs posted on the Internet raise censorship questions; network lets users reach millions. But what if messages are deceitful or slanderous?", *San Jose Mercury News*, May 1, 1994, page 1D(2). On sexism on the net, see Vaughan-Nichols, Steven J., "Connect Time: Online users should practice nonviolence and eschew sexism", *Computer Shopper*, Volume 14, Number 1, January 1994, page 668(2).

22. Both items are found in the "Electric Word" section in *Wired*, October 1994, on page 33.

23. Subjects were found to seek feedback more from a computer than another person, performance declined more when feedback was from another person rather than a computer, and self-esteem and public and private self-consciousness interacted with person-mediated feedback to negatively affect performance. See Kluger, Avraham N., and Seymour Adler, "Person- versus computer-mediated feedback", *Computers in Human Behavior*, Spring 1993, Volume 9(1), pages 1-16. For results contradicting previous findings that computer use results in deindividuation, see Matheson, Kimberly, and Erland Hjelmquist, "The impact of computer-mediated communication on self-awareness", *Computers in Human Behavior*, 1988, Volume 4(3), pages 221-233. On electronic alienation, see Kobielius, James, "Too much technology can lead to electronic alienation", *Network World*, Volume 7, Number 44, October 29, 1994, page 35.

24. From the "Electric Word" section of *Wired*, October 1994, page 37. The wording and description of the report itself reveal a joking attitude which does not appear to take the problem too seriously.

25. Greenfeld, Karl Taro, "The Incredibly Strange Mutant Creatures Who Rule the Universe of Alienated Japanese Zombie Computer Nerds", *Wired*, Premiere Issue, 1993, page 67.

26. Ibid., page 67.

27. From *The Oxford English Dictionary, Second Edition, Volume VII: Hat-Intervacuum*, prepared by John. A. Simpson and Edmund. S. C. Weiner, New

York: Claredon Press, and London, England: Oxford University Press, ©1993, page 1086.

28. Video game violence encourages neglect of responsible behavior, since violent actions are represented while consequences of those actions are not; diegetically speaking, "Game Over" is the worst thing that can happen to a player (admittedly, the stakes are slightly higher for arcade games, where a quarter or two might be at stake). The gap between actual consequences of violence and their representation on-screen is always greater than the gap between the representation of the interaction and the interaction itself. Situations with real-world consequences, such as Internet postings, use much the same interface hardware as games do, so it seems almost inevitable that phenomena like "flaming" would occur, as users neglect to consider the possible consequences of their actions. (For further discussion on the detachment of actions and consequences, see the section "Dis-integration and Reintegration" in Chapter 8.)

29. I am aware that making discrete distinctions and numbering them may appear to be a form of quantization; however, as I have pointed out, these are merely points of reference along a continuum, and are by no means as discrete as they might seem at first.

30. Stipp, David, "Some computers manage to fool people at the game of imitating human beings", *The Wall Street Journal*, November 11, 1991, page B58(W) and page B4C(E).

31. Morkes, John, "The Leprechaun Effect", *Wired*, January 1994, page 28.

32. Eastham, Kim, "Artifishal Experience", *Wired*, July 1994, page 122. For an article on interactive plant simulations, see Fraunfelder, Mark, "The Interactive Life of Plants", *Wired*, July 1994, page 35.

7.
The Metaphor of Cyberspace

In 1984, that Orwellian year of dystopia, William Gibson's award-winning novel *Neuromancer* appeared, envisioning a new technological dystopia and centering around the author's coined word, "cyberspace". Today "cyberspace" is used to include all forms of electronic communication, computer imagery, and everything else existing in what Michael Benedikt calls "The realm of pure information", describing it as,

> A new universe, a parallel universe created and sustained by the world's computers and communication lines. A world in which the global traffic of knowledge, secrets, measurements, indicators, entertainments, and alter-human agency takes on form: sights, sounds, presences never seen on the surface of the earth blossoming in a vast electronic night.[1]

The word itself is an abbreviation of "cybernetic space", the word "cybernetic" first used in 1948 by mathematician Norbert Wiener, who defined it as the theory and study of communication and control in living organisms or machines. *Cyberspace*, then, denotes a "space" of communications and control —two of the main uses of digital technology. To some extent, cyberspace is the pinnacle of digital technology, wherein the concept of the stored variable is expanded and multiplied into an entire "universe" of information.

Containing, as it does, on-line representations of those who use it and communicate through it, cyberspace is a metaphor, both spatially and socially. It can be thought of in spatial and social terms, both of which it relies upon for conceptual coherence, while at the same time questioning what it means to be "spatial" or "social". Like all metaphors, cyberspace is an abstraction, disengaging the social world from the physical world and the mind from the body. It can mimic the physical world in its design, or it can diverge into other imagined worlds. For the user, it is a world which one can look into but never be physically present in; and while looking into cyberspace, the user can forget —and often wants to forget— the real physical world around them.

While cyberspace may be the most technologically-based disengagement of the social world from the physical one, it is descended from an array of precursors, each of which developed their own metaphors and degrees of disengagement.

Degrees of Disengagement

The construction of miniature shared social worlds, and the role-playing that occurs in them, seems a natural enough part of human existence. One need only look at how children play, and how some of this play develops as they grow older. Games of pretend and make-believe, like "playing house", for example, are quite common among 5-year-olds. Children may even dress up while acting out their roles, using props, furniture, and so on. While stuffed animals may be put in the role of companions and the teacups at a tea party might be empty, the children physically act out their roles; but such play is soon replaced by the use of dolls or "action figures" as surrogate characters through which the miniature world can be experienced vicariously.[2] This immediately opens up the possibilities; everything in the miniature world is scaled down, like the dollhouse and its furniture, and space takes on metaphorical significance; a few square feet can become the entire lawn and yard surrounding a dollhouse, and a two-story building might only be a few feet high. Conceptually, space has changed. Other toys, such as Lego bricks, have even smaller people, and entire cities can be built on a tabletop. Time is also compressible during play; whole cities can arise, wars can be fought, and civilizations can develop over the course of a long Saturday afternoon.

Lego villages, Fisher-Price Little People, Barbie, G. I. Joe, and other action figures may be miniature, but they are all highly representational in design; whereas representational board games, like *Life* and *Monopoly*, are smaller still and represent a further level of abstraction (compare the hotels of *Monopoly* with those built in a Lego village). The board game also limits play, structuring it with rules and turns, and makes it teleological by giving players goals (in many cases, the completion of which will determine whether they are a "winner" or a "loser", or perhaps a high scorer). The board game, compared to other forms of play, has less physical action, relying more on conceptual representation and player interaction; while one could build with Lego or play with action figures alone, games almost always require at least two players. Often, in games with less physical action, social interaction becomes important.

In the 70s, open-ended role-playing games appeared on the market in which players created their own character personas from lists of attributes, and used them from game to game. TSR's *Dungeons & Dragons* (or *D&D*, as it is commonly known) is the largest and perhaps most well known of these games. Using graph paper, dice rolls, and lists of values representing characters' attributes, *D&D* player-characters exist in a completely quantized universe where even such things as dexterity, intelligence, and experience are expressed as numbers (not all that strange, perhaps, considering that I.Q. test scores are sometimes used in "real" life to represent a person's intelligence). Monsters have "hit points", and battles are represented as a series of dice rolls (just as in other board games, like *Risk*). *Dungeons & Dragons* is quite abstract; players do not even need to have any visual representation besides a description, and perhaps something to indicate their positions on the graph paper maps, in which one quarter inch square usually represent ten feet square in the dungeon being mapped.

Board games, and the other toys mentioned above, all involve play in the realm of the physical world, albeit to varying degrees. *Dungeons & Dragons* is more abstract, though it still uses pencil and graph paper. With the first digitally-based toys, video games, play enters the nonphysical realm; here the player's surrogate appears as a graphic on-screen, to be manipulated remotely by joystick or paddle, adding lack of physicality to miniaturization and abstraction. Just as in board games, the video game player's activities are somewhat more limited and are often quite goal-oriented, although some programs, like *Myst*, invite playfulness and contemplation. The video game has even moved into the territory of the board game, as games like *Dungeons & Dragons*, *Scrabble*, *Risk*, Chess, Checkers, and so on, and versions of solitaire and other card games are now all available on the computer. Not only did the video game replace physical action with its on-screen depiction of action, it also provided a "computer player" for many games, becoming *playmate* as well as plaything.

Inspired by *Dungeons & Dragons*, two Essex University students in England helped make-believe worlds make the transition into cyberspace in 1979, with the development of the first MUD (Multi-User Dungeon, or according to others, Multi-User Dimension), in which players from all around the world could phone in to a central database that controlled events and mediated game play, using text descriptions of events and objects, read and written by players as they typed in their actions. Just as *D&D* became popular, so did MUDs, for many of the same reasons. Howard Rheingold described the rapid growth of MUDding;

> By July 1992, there were more than 170 multi-user games on Internet, using nineteen different world-building languages. The most popular worlds have thousands of users. Richard Bartle, one of the fathers of MUDding, estimated one hundred thousand past and present MUDders worldwide by 1992. MUD researchers Pavel Curtis estimates twenty

> thousand active MUDders in 1992. The MUDding population is now far smaller than the populations of other parts of the Net, but it is growing fast, and spawning new forms at an impressive rate.[3]

Games have encouraged the growth of cyberspace, and have likely been an influence on the attitude of play so often found there. Cyberspace's ability to connect people electronically makes it the perfect "place" for socially shared "worlds" to exist.

Cyberspace's development can also be traced through another thread, that of the communication of ideas and the transmission of thoughts in the form of images and sound. Storytelling and books are perhaps the earliest models of shared mental worlds, the listeners or readers forming an audience. The same is true, on a much larger scale, of the narratives of history around which nations and cultures are formed, an idea explored by Benedict Anderson in his book, *Imagined Communities: Reflections on the Origin and Spread of Nationalism.*[4] But in this era of communication, increasing numbers of communication links are extending beyond national and cultural borders, weaving them together, and the cyberspace they collectively form is beginning to resemble a single global entity. In 1949, philosopher Pierre Teilhard de Chardin described just such an entity, which he called the "noosphere". The noosphere is a new layer that de Chardin saw as superimposed over the biosphere, a kind of "thinking layer", composed of thoughts and ideas, which makes human beings distinct from all other life on the planet. The noosphere was seen as a from of convergence through which all cultures will become unified into a single world culture. Cyberspace, and particularly computer networks like the Internet, allow individuals to tap into and contribute directly to the noosphere, and to experience it as a socially shared world. Whereas cyberspace is abstract compared to the physical world, it is concrete compared to de Chardin's noosphere. It allows experiences to be shared, and electronic communication technologies are gradually changing the nature of those experiences.

Electronic communications media have been weaving links between individuals separated by distance since the telegraph. While the telegraph made long-distance communication (almost) instantaneous, the telephone brought with it a wider bandwidth allowing it to reproduce the human voice. Commercial radio and television brought sound and image to a mass market, while short-wave radio and amateur short-wave television opened up similar possibilities to the individual. Of all of these, short-wave radio's many-to-many communications (and the global reach of its airwaves) most closely resembles that of the computer network, making it an interesting precursor of cyberspace.

On both short-wave radio and computer networks, participants often meet for the first time over the medium, and conversation between strangers is not unusual. For the sake of unique identification, participants are not often identified by their real names, but rather by call signs (on short-wave radio and TV) or user IDs (on computer networks). Unlike people's names which more than one person can share, call signs, and e-mail addresses are unique. Like an e-mail address, the call sign has a great importance to many ham radio operators, and it can even signify status. According to an article in the ham radio magazine *QST*;

> Walk down the aisle at any hamfest, look at the call sign badges of any of the passers-by, and you'll see just how important a ham's call sign is to him or her: It's printed in BIG letters... and his or her name is printed below much smaller, almost as an afterthought. It's the same on the air: Call signs first, and make sure to get 'em right, then name, the weather and the rig info. It's true: To most hams, that combination of letters and numbers means more than his or her given name![5]

As a representation of the user, the call sign can have significance which is not immediately obvious to the outsider;

> One-by-two call signs, such as W8IO or K4VX, are considered to be *preferred call signs*, because of the implied status of the holder as an "old-timer" in ham radio. But in pursuing the idea of preferred call signs further, we run into an anomaly: One-by-three call signs that begin with a W or K are also generally thought of as preferred call signs, but one-by-threes that begin with N are *not*, because they have come to indicate the newer generation of hams. . . Call signs with two characters in the prefix are also considered as being nonpreferred by many hams. . . The situation is somewhat amusing to the disinterested bystander, but for the hams involved, it is very serious business, indeed. It's often difficult to understand for nonhams or newer hams to understand the importance an amateur attaches to his call sign. Call signs are *at least* as important as names, and most amateurs want a short one that is easy for others to remember. Or one that reflects their initials or name. Or one that has a good sound (on phone) or good rhythm (on CW [in Morse code]), or *both.*[6]

Although it is not the same as having a player-character or alternate identity, the call sign and user ID can be a kind of persona which a person takes on when entering cyberspace.

Ham radio enthusiasts also constitute a community in a manner similar to the users of a computer network. Certain frequencies are more popular than others, and occasionally some will carry regularly scheduled announcements. Hamfests, radio clubs, and the exchange of cards acknowledging contacts are also ways of building community; as on the Internet, people who meet on-line or on-the-air may get together in person from time to time, but the main action takes place electronically in cyberspace.

All of these various technologies —telegraph, telephone, radio, television, short-wave radio and short-wave television— are being combined together in the computer through the use of digital technology, opening up new possibilities for representation and interaction, and together constituting what we think of as cyberspace, in which the computer has so central a role. Through the concept of the electronically-stored variable, the computer can store and retrieve messages, unlike the other media listed above. The fact that storage and retrieval take place within the computer means that cyberspace is more reliant on a machine than shared worlds of the past were, for it is, to a large extent, stored in machines. There is more mediation and abstraction, considering the convoluted route that information takes from sender to receiver: communication is constructed in text, image, or sound, and then entered, scanned, or recorded into a computer; there it takes on machine-readable form, and is stored as a series of ones and zeroes; and *those* are turned into a modulated signal —and then the whole process is reversed at the receiving end of the line.

The construction of miniature shared social worlds and the communication that occurs within them always require some degree of disengagement, and along with it, metaphors to structure them and give them coherence, to counter the inevitable abstraction. The more intuitively an electronic world is designed to imitate something already familiar to the user, the more successful it will be, and easier to use. Two main metaphors are needed; spatial ones, creating a 'space' that participants are 'in' while communicating, and social ones, to create a sense of community (or *virtual* community).

Spatial Structures

The notion of *space* has changed greatly throughout history, with each new conception relying on prior models, to define what it was and what it wasn't. First, there is physical space; the space all around us, in which matter and energy are said to exist. From the early notions of a simple, absolute Euclidean space (visualized in Renaissance perspective) to the increasingly complex theoretical models that have replaced it, each version has come with its own philosophical implications. As space was mapped and divided by Cartesian grids, and techniques like calculus and analytic geometry grew, mathematics became a way of talking about space, and calculations grew more complex and more difficult (if even possible) to do by hand. Computers carried algebraic and geometric ideas even further, and today the "space" of cyberspace can be used to visualize various mathematical entities and models of space.

The need for a computing machine arose not only from the number and the extent of the calculations involved, but through changes in the way space was conceptualized. As Michael Benedikt writes;

> Insubstantial and invisible, space is yet somehow *there*, and *here*, penetrating, and all around us. Space, for most of us, hovers between ordinary, physical existence and something other. Thus it alternates in our minds between the analyzable and the absolutely given.
>
> Or so it was until modern physics and mathematics revealed space's anatomy, as it were, showing its inextricability from the sinews of time and light, from the stresses of mass and gravity, and from the nature of knowing itself. The early part of the twentieth century saw post-Euclidean geometry and the Theory of General Relativity admit the concepts of curvature and higher dimensions, introducing "inertial frames," "manifolds," "local coordinate systems," and "space-time" to all informed discourse about space. These ideas had myriad practical consequences. Physical space, we learned, is not passive but dynamic, not simple but complex, not empty but full. Geometry was once again the most fundamental science.[7]

Many post-Euclidean models of space appear counterintuitive, since their departures from the Euclidean model involve velocities and spatial structures larger or smaller than those we are accustomed to dealing with in everyday life. Because of this, post-Euclidean models seem to be more abstract, even if they do, in the end, produce a more consistent picture of the physical universe.

The fact that there are different ways to conceptualize configurations of space produces questions as to how spaces can differ, and leads to the idea of the dimension, which provides some measure or description of space. (I have already mentioned in chapter three how dimensionality becomes flexible in digital media, and thus in cyberspace.) In his essay "Cyberspace: Some Proposals", Michael Benedikt takes an in-depth look at space and dimensionality, and the variety of options in the visual design of cyberspace. For instance, he writes of *extrinsic* and *intrinsic* dimensions of objects in cyberspace;

> In Euclidean geometry, a "point" has no character: it has no size and no intrinsic, inherent properties. When one makes the statement "There *exists* a point such that . . .," the point exists as pure position, and herein precisely lies its conceptual usefulness. But in the physical

> world (and certainly in the world of actual representation) a point is always a "something"—a dot, a spot, a particle, or a patch of a field—a *point-object* to which one or more values can be attached that are descriptive of its character. In other words, unlike a true, Euclidean point, a point-object might have a color, a shape, a frequency of vibration, a weight, size, momentum, spin, or charge—some intrinsic quality or set of qualities that is logically (though it may or may not as matter of empirical fact *be*) independent of its position in space.
>
> Now, any **N**-dimensional state or behavior of a system can be represented in what I call a data space of point-objects having **n** spatiotemporally locating or *extrinsic* dimensions, and **m** *intrinsic* dimensions, so called because they are coded into the intrinsic character of the point object.[8]

The notion of a "dimension", then, is broadened into a that of *a direction of measurement*, a value along a continuum, and "dimension" used in reference to spatiotemporal measurements becomes a subset of the broader definition. "Space", too, comes to mean more than physical space, but rather a realm in which multiple possibilities or permutations exist. With this broadening of the definition, "space" becomes more than that which is experienced, but also something that can be imagined in different configurations (such as phase space).[9]

"Space" is also used to describe computer memory; we talk of whether or not a program or file will "fit" in memory, of a Ram "cache", or a file "size". There is some correlation between the two, as computer memory does take up physical space (only so many memory cards or SIMMs can fit in any given computer model). Even the notions of *in*put, *through*put, and *out*put reflect the spatial metaphor. Since the invention of the stored-program computer there has been a metaphor of a "space inside the computer", in which one could "store" and later "retrieve" a variable in memory, as if it were a physical object. (One might even ask how computer memory space is related to cyberspace: how much cyberspace can be created from a given number of bits? If text-based worlds are included, however, these questions may not really have a measurable, quantifiable answer, since one would first have to quantize cyberspace into elemental units.) Despite the fact that human memory supplies the metaphor for computer "memory", no such metaphor of space has ever been used to describe *human* memory; the notion of space applied in one case but not the other underlines the differences between human and computer memory. The computer's memory had, at first, a visible analog in the banks of blinking lights, indicating the states of its variables; input and output were clearly defined, simplified to numeric form; and the functioning of the computer was limited to the instructions it was given, unlike the automatic and partially uncontrolled nature of human recall. And unlike the brain, the computer's memory had rigid and precisely defined limits as to how many variables or bits it could "hold" at once; this more than anything was responsible for the metaphor, for the notion of "bits" quantized memory and made it a measurable quantity, by which comparisons and measurements could be made. The brain's memory could never be "full", but a computer memory could.

Although human memory may not be referred to in spatial terms, spatial metaphors are ingrained in forms of thinking. "Cognitive mapping" and visualization in the "mind's eye" can form what one might call cognitive space. Just as cyberspace is the place where electronic communication "takes place", cognitive space could be said to be where thought or dreaming take place, or where spaces are reconstructed from memory. Computer-generated scientific visualization is a kind of concretization of cognitive space, since it is used to show what would have otherwise had to be described and imagined, giving it a precision not possible otherwise. In one sense, cyberspace itself could be seen as a kind of shared cognitive space —a space of communications that underlies the shared social worlds described above.

Cyberspace, as a space of communications, is by definition a social space. People are seen as *points* (as when one speaks of point-to-point connections), communication links are referred to as *lines*, and sets of such lines are networks; a geometric metaphor is carried throughout, culminating in the *space* of cyberspace. Cyberspace itself is undergoing renovations in its spatial metaphors. MUDs began as worlds completely composed of text, but around 1986, Lucasfilm developed *Habitat*, a MUD designed for QuantumLink, an on-line service for owners of Commodore 64 computers. Combining the idea of the MUD with cartoonlike video game graphics, Habitat was the first MUD with visual representation of its players and environment. According to Chip Morningstar and F. Randall Farmer, the two main architects of Habitat,

> The largest part of the screen is devoted to the graphics display. This is an animated view of the player's current location in the Habitat world. The scene consists of various objects arrayed on the screen, such as the houses and tree. The players are represented by animated figures that we call "Avatars." Avatars are usually, though not exclusively, humanoid in appearance. . . Avatars can move around, pick up, put down, and manipulate objects, talk to each other, and gesture, each under the control of an individual player. Control is through the joystick, which enables the player to point at things and issue commands. Talking is accomplished by typing on the keyboard.

> The text that a player types is displayed over his or her Avatar's head in a cartoon-style "word balloon."[10]

Habitat space, like a video game, was made up of adjoining screens through which players' Avatars can move. Its basis in object-oriented data representation allowed players to carry and use a wide assortment of objects, weapons, tools, and even spend tokens, Habitat's currency. Like the text-based MUDs, players must be in the same region as those they are chatting with, but the feeling of off-screen space is much stronger since spaces *can* be represented on-screen (in a text-based MUD, all space is "off-screen", if the word can even be said to apply in a text-based world).

Following these are virtual reality-based MUDs, in which participants appear and move through representations of three-dimensional spaces. Multi-user virtual reality was first available to the public in the form of games, such as the games made by Virtual World Entertainment; in *Battletech* and *Red Planet*, in which up to eight players play at once, and can see each other's vehicles on-screen. At lower resolutions, there are maze-like games available on CD-ROM, like *Doom*, which can be played by multiple players over a network.

The spatial structures used in the design of cyberspace affect the social structures built within them. For example, in Habitat, players' Avatars were visible on-screen and more limited in their appearance —and thus more similar to each other— than the characters in text-based MUDs who are described verbally and can be anything from a human being to a gelatinous cloud. Likewise, the scenery in Habitat is more limited, but at the same time more concrete; players all see the same graphics when in a particular area, unlike text descriptions which are more ambiguous. On the other hand, graphic representations allow for greater simultaneity than text; several players can enter a space at once and react to each other immediately, as opposed to having a series of text sentences announcing their arrivals appear on-screen before any actions can take place.

The spatial metaphors structuring cyberspace are the ground upon which the social metaphors structuring cyberspace are built. These social structures are often the main reasons for cyberspace's existence, and like spatial metaphors, they can be built and manipulated according to the desires of users, though they are limited by the technologies used. Together, both forces make these electronically shared communities different from social structures in the material world.

Social Structures

Consumer-produced alternative mass media like short-wave radio and TV, or websites on the World Wide Web, give witness to the transition that is occurring in electronic media and the users it addresses. We might state it as a shift from *audience* to *community*, as one-to-many communications give way to many-to-many communication. The former was present in theater, commercial radio, television, and film, and print media like newspapers and magazines, which were largely one-way and had limited means of feedback (letters to the editor, for example, could themselves be edited, or ignored and left out entirely). There was a source from which transmissions came, and an audience who received them; although readings and interpretations varied among audience members, their was little they could do to respond to the programs they received. In the community structure, transmissions are shared among all (or potentially all) of those involved, with no distinction made between performers and audience. There is no central stage upon which all events occur, and it is no longer possible to see all the action that is occurring, due to the increased number of participants. While community-structured media does not replace audience-structured media, it displaces it, competing for a person's time, and as a venue for news, entertainment, education, and so forth.

These changes are manifest in both the individual and in the community as a whole. First, the individual has a new role, as a participant rather than mere observer, and finds himself or herself dealing with others in a similar role. This is quite different from the former audience-based media, in which all of the people or characters observed by the audience were fictional characters played by actors, or were as imaginary as characters in fiction. To an observer accustomed only to audience-based media, there is a clear distinction between fictional characters (who always appear in mediated form) and real persons, those friends and associates in the material world, who exist in physical form. One can interact with real persons, and this interaction often requires the expression of an identity (in one's replies and reactions), whereas fictional characters are merely observed, requiring no reaction or response from the observer.

But while fictional characters are limited to media, real people are not completely excluded from it; real people appear in pictures and written accounts. At first, there were only books and paintings and oral accounts of events; but when mass media appeared, events were reported daily in newspapers, radio, newsreels, television, and so on; real people appeared in mediated form more than ever before. There were, however, still distinctions made (usually) between nonfiction depicting actual events, and fiction wherein actors played characters. Today such distinctions have been thoroughly questioned; there is much more realization of the constructedness and bias inherent in all mediated events, and a greater acknowledgement of the gaps between the media representation and the referents being represented.

Called upon to be participants in a mediated world, it would seem natural that people would follow the model of the actor from audience-based media, and want to create an alternate identity (especially if they think others will be doing so). Here again, the alternate identity will inevitably overlap to some degree with what the person thinks of as their "real" identity, since the alternate identity, in the end, is still created and controlled by the person in question; even fantasy characters have to come from somewhere.

The alternate identity can be either an overt one, in which everyone is aware of the "acting" going on; or a covert one in which deception based on false identities occurs. In the former, characters are usually involved in games or game-like situations, fantasy MUDs where players are wizards, warriors, and so on, and everyone openly presents the mask they are wearing; a player is represented by a player-character. To a lesser extent, the more social MUDs and non-goal-oriented MUDs are similar; players are identified by aliases, and their appearances or descriptions are often bizarre and clearly fictional, although their behavior and discussion are more than simply that of a fantasy world. Here too, it is clear that people are deliberately using fanciful personae unlike their appearances in the physical world. However, in other areas of cyberspace, like the public and private conferences and discussions of serious topics, systems like the WELL (Whole Earth 'Lectronic Link), and corporate e-mail, users are expected to represent themselves in an honest and realistic fashion, as they would in face-to-face encounters.

In most non-game conferencing, users are expected to act responsibly, be accountable for what they say, and not deliberately misrepresent themselves. User IDs can function in place of names, but there are usually "finger" commands which reveal users' real names. Some systems are even designed to discourage or prevent anonymity. The WELL, for example, requires users to attach their user ID to their postings, so that nobody is anonymous (even if pseudonyms are used, it is possible to trace the user's real user ID). But most systems allow anonymity, and there is plenty of opportunity for covert role-playing. Age and gender deception are often reasons for creating a false identity on a computer network. While there are those who misrepresent their gender for what they see as the power trip involved in fooling another person, many people who do so are curious and want to experience, at least on-line, what it would be like to be the opposite sex (though most would never dream of doing so in face-to-face encounters). Sometimes, the experiment can have enlightening results; in one case, a male user masquerading as a woman was upset by the aggression and vulgarity of other men on the chat lines.[11] Many chat rooms have had so much masquerading that no one takes anyone or anything seriously any more, and they have practically become gaming MUDs. But when too many people are lying and no one is believed, deceptions are not as effective and users looking for a thrill may be tempted to move to more serious conferences, where damage can be done and people can be hurt. In *The Virtual Community*, Howard Rheingold writes of Lindsy Van Gelder, and an experience on the CB channel of CompuServe with a user known as Joan;

> After Van Gelder encountered Joan in a wide-open, public chat session, she engaged her in private chat. She learned that Joan was a neuropsychologist, in her late twenties, living in New York, who had been disfigured, crippled and left mute by an automobile accident at the hands of a drunken driver. Joan's mentor had given her a computer, modem, and subscription to CompuServe, where Joan instantly blossomed. . . . Joan connected with people in a special way, achieved intimacy rapidly, and gave much valuable advice and support to many others, especially disabled women. She changed people's lives. So it was a shock to the CB community when Joan was unmasked as someone who in real life, IRL, was neither disabled, disfigured, mute, nor female. Joan was a New York psychiatrist, Alex, who had become obsessed with his own experiments in being treated as a female and participating in female friendships.
>
> The sense of outrage that followed the revelation of Joan's identity came first from the direct assault on personal relationships between Joan and others, friendships that had achieved deep intimacy based on utter deception. But the indirect assault on the sense of trust essential to any group that thinks of itself as a community, was another betrayal. Van Gelder put it this way; "Even those who barely knew Joan felt implicated—and somehow betrayed—by Alex's deception. Many of us online believe that we're a utopian community of the future, and Alex's experiment proved to us all that technology is no shield against deceit. We lost our innocence, if not our faith."[12]

Elsewhere, Allucquere Rosanne Stone relates a similar story, about a man masquerading as an older disabled woman for years who was unmasked. She quotes one woman whose reaction to the deception was "I felt raped. I felt that my deepest secrets had been violated."[13] Such predatory behavior on the unsuspecting is a rape of the mind, and a deception which can be as harmful as any occurring off-line. Or worse, since there is not yet any legal recourse for the victim to take action against her deceiver.

Due to employment quotas and other programs favoring certain identities over others, some identities can be more profitable than others, and gender deception can even take place on a corporate scale. According to the *Los Angeles Times*, Interactive America Corporation was investigated by the FCC in 1994 for possibly misrepresenting itself as a female-owned business, which it did in order to qualify for a discount during an auction process; the company bid $17 million for 15 licenses, and received a $3.3 million discount because of its gender classification.[14]

Computer law is slowly developing, as new cases bring discrepancies and legal questions to light, and cyberspace is gradually coming under jurisdiction of the court system. One such case appeared in "Dispute over computer messages: free speech or sexual harassment?" on the front page of a 1994 issue of *The New York Times*;

> In a case that highlights the tension between free speech and illegal sexual harassment—and raises new questions as to what legal protections apply to computer messages—a junior college in California has agreed to pay three students $15,000 each to settle charges stemming from its men-only and women-only computer conferences.
>
> Santa Rosa Junior College reached the settlement after the Department of Education's Office for Civil Rights, with whom the claim was filed, told the college in a letter last June that it had found a probable violation of the Federal law prohibiting sex discrimination in schools.
>
> Two plaintiffs in the case were women who had been the subject of anatomically explicit and derogatory remarks posted on the men-only computer conference. . . . The third plaintiff, Dylan Humphrey, saw the computer messages and told the women about them, and said he suffered retaliation by the college for his actions. "I thought it was best that if someone was being spoken about, they should have a chance to deal with it," said Mr. Humphrey, adding that he was very pleased with the outcome of the case.
>
> Legally, the case presents several firsts. The Office for Civil Rights says there have been no prior legal rulings on how much First Amendment protection should be accorded to messages on college computer bulletin boards. And computer law experts say that they know of no previous charges of sexual discrimination arising out of single-sex computer services.[15]

Computer bulletin boards and MUDs are disengaged from the physical world, but not disconnected; what goes on in cyberspace has an effect on the rest of the world. An on-line persona can affect the rest of a person's life when a user identifies too strongly with it. As Howard Rheingold points out, referring to the work of social scientist Kenneth Gergin,

> It is self-evident that many of us communicate with many more people every day, via telephone, fax, and e-mail, than our great-grandparents communicated with in a month, year, or lifetime. According to Gergin, social saturation is an effect of internalizing parts of more people than any humans have ever internalized before. Our selves have been "populated" by many others, Gergin claims.[16]

As the story of "Joan" shows, you must be careful as to who "populates" your psyche.

The alternate identity, disengaged to a degree from one's person, is seen by some as freeing them from the responsibility they would otherwise have for their acts. This raises questions of ethics in cyberspace; in an article entitled "A Rape in Cyberspace", Julian Dibbell describes events (or rather, the text describing events) which occurred in LambdaMOO, a text-based MUD. One of the MUD characters created a "voodoo doll" through which he could make other characters do horrible things against their will, including rape. He was only stopped later by another character who had a gun which, when fired, enveloped its target in an impermeable cage. After describing the grisly events in graphic detail, Dibbell writes;

> These particulars, as I said, are unambiguous. But they are far from simple, for the simple reason that every set of facts in virtual reality (or VR, as the locals abbreviate it) is shadowed by a second, complicated set: the "real-life" facts. And while a certain tension invariably buzzes in the gap between the hard, prosaic RL [real life] facts and their more fluid, dreamy VR counterparts, the dissonance in the Bungle case is striking. No hideous clowns or trickster spirits appear in the RL version of the incident, no voodoo dolls or wizard guns, indeed no rape at all as any RL court of law has yet defined it. The actors in the drama were university students for the most part, and they sat undramatically before computer screens the entire time, their only actions a spidery flitting of fingers across standard QWERTY keyboards. No bodies touched.[17]

Dibbell writes that one of the victims was brought to tears as she sat at the keyboard; and many players agreed that the violations were serious. The offending character was later killed off, an unusual and rare act in this particular MUD, but most players agreed that it be done. A short time later, however, a new character, strangely familiar in behavior if not description, appeared in LambdaMOO; the person behind the character had obtained a new Internet account and was logging back on.

Cyberspace's similarity to a video game makes it all too easy to forget the links that exist between characters or personae and the real-world people behind them. Alternate identities help to disengage the two spheres, but connections between them remain, and thus questions of ethics and morals can be applied to cyberspace and the actions occurring there. Hackers and computer criminals have gone to prison for deeds done on-line, and people's lives are strongly influenced and affected by what goes on in cyberspace.

On the other hand, on-line communities and communities in the material world are not the same thing; an on-line community, no matter how close knit, is still an abstraction to some degree, and its effects on the material world —and individuals— can be detrimental. Erik Davis writes;

> MUDs are spectacles that bind. After *Wired* magazine ran Howard Rheingold and Kevin Kelly's breezy, upbeat article on these virtual realities, horror stories swamped the letters page: students having to be forcibly removed from their workstations, academic careers run aground, relationships destroyed. Having spent half of last summer stuck in a MUD, I have great respect for these sobering tales. After I caught the MUD bug, I'd hang out online until my eyes fried, chatting with excellent pals, tinkering with my room, hazing newbies. My modem was my lifeline; material friends grew tired of my constant busy signal and ceased calling. Increasingly unwashed, subsisting on toast, chips, and other goods easily consumed at the keyboard, I let my body wilt in the summer heat. In that six-week period, my freelance career nearly ground to a halt, and only sheer necessity and a novel degree of will on my part pushed me back into the cold turkey that MUDers call RL: real life.[18]

Similar patterns have occurred in compulsive video game playing, in which players forego the physical world for self-contained, highly-controlled on-line worlds (for example, people who play computer solitaire for hours on end might only rarely (or never) take out a deck of cards to physically play it). And if some people find the games irresistible, adding the element of communication with other players makes MUDs and on-line communities even more attractive, especially when combined with alternative identity creation possibilities and 24-hour availability.

As we have seen with the *otakus* of Japan, some people participate to the point of communication addiction. This problem affects so many people that there are even therapy groups forming on the Internet (ironically!). In *The Virtual Community*, Howard Rheingold described a Usenet newsgroup, alt.irc.recovery, for people addicted to Internet Relay Chat. He writes;

> During informal monitoring of the alt.irc.recovery newsgroups, I noticed frequent confessions that constant IRC use led chat addicts to spend less time with physical friends in the immediate geographic neighborhood; in the long run, this reclusiveness led to loneliness even more acute when they weren't logged onto the one community in which things went smoothly. Like the question of MUD addiction, any question about proper ways to use a communications medium leads to deeper questions about what our society considers to be proper ways to spend time.[19]

Elsewhere in the book, Rheingold describes a conversation he had with Michel Landaret, the man in charge of a French chat system. Landaret reported that within a thirty day period (720 hours), one on-line addict spent 520 hours on-line; the maximum number of consecutive hours that a user spent on-line was 74; and the highest phone bill for a two month period was over $25,000.[20]

Communication addiction is certainly a danger for people who have little else in their lives, but even those who are not addicted can find more and more of their time being used on-line. Many business and mobile office workers will have to spend more time on-line whether they like it or not. And many people who are otherwise difficult to reach may have to be reached on-line, for example, through e-mail.

Although virtual communities supplement or *displace* conventional communities, they should not be allowed to *replace* them. Certainly, computer-mediated communication has its positive side; but while it may be good for physically handicapped and disabled people, for elderly who don't go out, or socially shy and isolated individuals, it does not solve the problems they face, but rather provides others an excuse to not deal with these people and their needs in person. And other groups, such as the mentally handicapped or the illiterate, have even less opportunity to participate on-line since computer-mediated communication is almost wholly the result of mental and verbal skills alone. Although there are computers designed for use by the blind, spoken communication is still easier and faster to use than on-screen text. Shy people may be able to communicate more freely on-line, but, as in the above examples, this may have the effect of keeping them away from social situations in the material world, allowing their limited social skills to deteriorate even further. There are other situations in which the computer cannot reproduce in-person communications in the face-to-face mode; two people who do not speak the same language can communicate through gestures and so on in person, but they would be unable to connect in an all-text network connection. And, finally, there are the nontechnical, computer illiterate people who simply are not on-line. Even if everyone *wanted* to be on-line there would always be those who have never had the opportunity due to lack of education or access, or simply due to poverty.

In virtual communities as in physical ones, money is still a determinant. Some people are amazed at the economic stratification of Los Angeles; the poverty of south central areas like Watts, compared to the "gated" communities like Bel Air, where even the neighborhood streets are off limits to outsiders. What is often forgotten is that all virtual communities are "gated" communities, and most charge admission as well. There are a few, however, like Santa Monica's PEN (Public Electronic Network) system, open to all, which even provide free public access, even to the homeless. But larger providers, like America On-Line, Netscape, and others, charge a monthly rate (not to mention the cost of the equipment needed for logging on).

Questions as to who has access to cyberspace and who doesn't also extend to an international level, where the cyberspace race continues at uneven rates. While the "electronic frontier" is being tamed, it may be the case that another, new form of empire-building is going on. In his book *Electronic Colonialism: The Future of International Broadcasting and Communication*, Thomas L. McPhail writes,

> . . . two major changes occurred during the late 1950s and early 1960s which have set the stage for the fourth and current era of empire expansion.
>
> The two major changes are the rise of nationalism, centered mainly in the Third World, and the shift to a service-based economy in the West which relies substantially on telecommunications systems, where traditionally geographical borders and barriers to international communications are being rendered obsolete. The postindustrial society, with information-related services being the cornerstone, has significant implications for industrial and nonindustrial nations alike. Military and mercantile colonialism of the past may be replaced by "electronic colonialism" in the future. A nation-state may now be able to go from the Stone Age to the Information Age without having passed through the intervening steps of industrialization.
>
> Electronic colonialism is the dependency relationship established by the importation of communication hardware, foreign-produced software, along with engineers, technicians, and related information protocols, that vicariously establish a set of foreign norms, values, and expectations which, in varying degrees, may alter the domestic cultures and socialization processes.[22]

For all the social and spatial metaphors used, cyberspace is neither a society nor a country, but the product of hardware and software; it is not discovered and settled but created and shaped. Rather than coming across land and then deciding what to build on it, cyberspace comes into being with a use already in mind for it —and only then. Cyberspace itself is not colonizable, because it is always already colonized, for how can uncolonized or unexplored cyberspace be said to exist in the first place? The real colonization going on occurs just outside of cyberspace, in the displacement —or replacement— of traditional cultures, communities, and social structures.

The quantization of everyday life helped to prepare the conceptual framework that houses cyberspace, a "world" which is not only completely quantized, but is made up entirely of bits. The growth and use of cyberspace will, in turn, continue to further this quantization, while the implementation of new and planned technologies will create a new environment around us in the physical world; together, these forces change are changing perception, representation, and cognition itself.

NOTES

1. The quotes are taken from pages 3 and 1, respectively, of Michael Benedikt's "Introduction" in *Cyberspace, First Steps*, edited by Michael Benedikt, Cambridge, Massachusetts, and London, England: The MIT Press, ©1991.
2. Games of make-believe wherein players act out their parts, even to the point of dressing up, are also played by adults, in everything from commercial role-playing games like *How to Host a Murder*, to paint-ball wars and laser tag, to the annual get-togethers of the "mad masters" in Jean Rouch's ethnographic film *La Maitre Fous* (1950).
3. Rheingold, Howard, *The Virtual Community: Homesteading on the Electronic Frontier*, New York: HarperCollins Publishers, ©1993, page 145-146.
4. Anderson, Benedict Richard O'Gorman, *Imagined Communities: Reflections on the Origin and Spread of Nationalism*, London and New York: Verso, ©1991.
5. Palm, Rick, "Back to the Future: Pick Your Own Call Sign?", *QST*, February 1994, page 84.
6. Sager, Phil, and Rick Palm, "An Overview of Amateur Call Signs—Past and Present", *QST*, May 1994, page 54.
7. Benedikt, Michael, "Cyberspace: Some Proposals", in Benedikt, Michael, editor, *Cyberspace: First Steps*, Cambridge, Massachusetts, and London, England: The MIT Press, ©1991, page 125.
8. Ibid., pages 134-135.

9. For an explanation of phase space, see footnote 38 of chapter 3.
10. Morningstar, Chip, and F. Randall Farmer, "The Lessons of Lucasfilm's Habitat" in Benedikt, Michael, editor, *Cyberspace: First Steps*, Cambridge, Massachusetts, and London, England: The MIT Press, ©1991, page 275.
11. Poole, Gary Andrew, "Gender bending on the net", *Open Computing*, Volume 11, Number 5, May 1994, page 128.
12. Rheingold, Howard, *The Virtual Community: Homesteading on the Electronic Frontier*, page 165.
13. Stone, Allucquere Rosanne, "*Will the Real Body Please Stand Up?:* Boundary Stories about Virtual Cultures", in Benedikt, Michael, editor, *Cyberspace: First Steps*, Cambridge, Massachusetts, and London, England: The MIT Press, ©1991, page 83.
14. Apodaca, Patrice, "Sun Valley firm, others lose interactive rights after failing to pay up", *Los Angeles Times*, Volume 113, October 8, 1994, page D1.
15. Lewin, Tamar, "Dispute over computer messages: free speech or sexual harassment?", *The New York Times*, Volume 144, September 22, 1994, page A1 and D22.
16. Rheingold, Howard, *The Virtual Community: Homesteading on the Electronic Frontier*, page 169.
17. Dibbell, Julian, "A Rape in Cyberspace", *Village Voice*, December 21, 1993, page 36.
18. Davis, Erik, "It's a MUD, MUD, MUD, MUD World", *Village Voice*, February 22, 1994, page 42.
19. Rheingold, Howard, *The Virtual Community: Homesteading on the Electronic Frontier*, page 182.
20. Ibid., pages 228-229.
21. Ibid., page 33.
22. McPhail, Thomas L., *Electronic Colonialism: The Future of International Broadcasting and Communication*, Beverly Hills, California, and London, England: SAGE Publications, Inc., ©1981, pages 19-20.

IV.

Perception

Representation

Cognition

8.
The Digital Environment

Mass media, machine mediation, and notions of cyberspace are all examples of how human beings are losing touch with the natural world, an idea found in the writings of O. B. Hardison and Neil Postman as well as the films of Godfrey Reggio. In our technological society, many people have little contact with nature, leaving their homes on sidewalk or asphalt, riding in cars to work in a building all day and back again at night, having spent most of their time in manufactured environments. There is an effort to keep the two domains separate; signs tell us to "Keep off the Grass", and "dirty" has come to have negative connotations.

Chapter one described how the natural world came to be thought of in quantized, measurable terms; the taming of the analog frontier. Today, our environment is finally starting to catch up —thanks to technology— to our quantized view of the world. Once perception is thoroughly quantized, it becomes a force, through which representations become quantized, turned into measurable quantities that can be specified to varying degrees of precision —and so the system becomes self-justifying. Communication is closely related to perception; it is that set of one's perceptions of other people and what they are saying or doing (as well as one's own perceptions of how one's own self is perceived by others). As communication becomes technology-dependent, so does perception. Since the invention of the telescope and the microscope, technology has been extending the senses of sight and sound; Marshall McLuhan even went so far as to describe electronic technology as an extension of the human nervous system. But technology not only provides the instruments through which we sense the world, it physically alters the places where it is in use. Just as the chairs in a room containing a television will often be set facing the television, technology is a deciding factor in the design of the environment, and in some ways, technology has *become* the environment. Much of our surroundings is carefully monitored and controlled by technology, when one considers air conditioners, heaters, and temperature and humidity control, and so on. Artificial lighting inside buildings often supplements natural lighting or replaces it entirely. Computer rooms are typically cut off from natural lighting;

one is more likely to encounter Microsoft *Windows* than conventional glass ones.

Digital technology affects its physical environment, and its effects are a result of both the nature of the technology itself and the uses to which it is put. The next three sections trace a trajectory arising from these effects. The first deals with the way digital technology reconfigures its surroundings, taking them apart and reassembling them in ways that give technology an important role. The second is concerned with the user's point of view, and looks at how digital technology advances media's engulfment of its audience, gradually drawing more of its time and attention, and providing more of the viewer's contact with or knowledge of the outside world. Finally, the last section combines elements of the first two, examining how technology has become abstracted and invisible, blending into and becoming a part of the environment itself, until it is merely a given, taken for granted and yet wholly relied upon.

Dis-integration and Reintegration

As discussed earlier, the usefulness of quantization is the concept of divisions breaking something into discrete, separable, component parts. These parts can be studied individually from the whole and can be categorized. The assigning of numerical values and levels turns quality into quantity, and these quantities can then be compared and measured. Conceptually, wholes are divided up or taken apart, dis-integrated into component pieces. They may be reintegrated, but in a way that reflects the understanding of those pieces at the time of their disassembly; the way the functions of individual parts of a whole are seen depends on the way the whole is divided into parts. Different divisions result in different views of the whole, and examples of this can be seen across different cultures, wherein things in the physical world, as well as conceptual and philosophical ideas, are divided into different categories and combined in varying ways, resulting in differing world-views. But supposing one subscribes to the idea that for any given whole, there is always "one best way" to subdivide it into parts?

Philosopher of technology Jacques Ellul writes at length of what he calls *technique*, a term which includes both technology and the uses to which it is put, and even the standardized methods used for attaining a particular objective. The application of technique is premeditated, and sees efficiency as a way of evaluating a method, giving rise to the notion of "one best way" that maximizes efficiency. Since *technique* has efficiency as its end, Ellul points out, any choice between means is reduced to one: the most efficient. Ellul examines how technique has been applied throughout history, and in various institutions of society. He points out, for example, how technique is opposed to liberalism, and how it turns democracy into dictatorship, by placing one means above all others, eliminating choice. Thus technique can come to dictate human activity, through the desire for efficiency, "And here reason appears clearly in the guise of technique."[1]

Through digital technology, *technique* is no longer limited to human beings, but can be carried out by computers, which are often used to solve problems dealing with the determining of maximum efficiency. Ellul foresaw the rise of the computer, and the subjugation of human beings to the efficiency of the machine;

> Now that statistical operations are carried out by perforated-card machines instead of human beings, they have become exact. Machines no longer perform merely gross operations. They perform a whole complex of subtle ones as well. And before long—what with the electronic brain—they will attain an intellectual power of which man is incapable. . . . However, there are spheres in which it is impossible to eliminate human influence. The autonomy of technique then develops in another direction. Technique is not, for example, autonomous in respect to clock time. Machines, like abstract technical laws, are subject to the law of speed, and co-ordination presupposes time adjustment. . . . Technique obeys its own specific laws, as every machine obeys laws. Each element of the technical complex follows certain laws determined by its relations with the other elements, and these laws are internal to the system and in no way influenced by external factors. It is not a question of causing the human being to disappear, but of making him capitulate, of inducing him to accommodate himself to techniques and not to experience personal feelings and reactions.
>
> No technique is possible when men are free. When technique enters into the realm of social life, it collides ceaselessly with the human being to the degree that the combination of man and technique is unavoidable, and that technical action necessarily results in a predetermined result. Technique requires predictability and, no less, exactness of prediction.[2]

Through quantization and digitization, digital technology contributes to the power of technique not only by allowing measurements and comparisons of measurements to be made for whatever is being represented digitally, but to perform these comparisons effortlessly, millions of times per second, on computers. Reexamined in light of digital technology, Ellul's *technique* has become more powerful and concrete in the algorithmic behavior of the computer, which is now used to make many social systems and bureaucratic institutions more algorithmic in the way *they* are run. Technique dis-integrates them, and reintegrates them, attempting to conform them to the notions of efficiency held by those in power. The very technology that provides the greatest flexibility, then, can also be used to promote and justify the most rigid of structures.

Dis-integration and reintegration occur in other social planes as well; the last two chapters have described ways in which social groups and group interactions have been broken up into individual machine-mediated interactions, reintegrated into networks, MUDs, and cyberspace. Offices have dis-integrated into cadres of mobile works, while the office has been reintegrated into the home, and the work day extended to all hours of the night. "Work", for many, involves not so much physical labor, but rather button-pushing, machine-tending, and message-sending. Communication itself has been broken down into its constituent parts; messages are sent, stored, and retrieved at a later time, and the message itself is often reduced to the barest essentials necessary to convey its meaning. With work separated from physical labor, exercise and fitness clubs have become popular, to fill in the missing physical activity. Rather than integrating work and exercise together, many people spend hours of mental activity that are physically inactive, followed by shorter periods of physical activity, 'workouts' which accomplish no labor or work apart from the upkeep of the body.

So much of contemporary life has become dis-integrated into a series of functionless pleasures and pleasureless functions; aspects of enjoyment and utility have become separated. In the same manner that work and exercise have become detached, so have food and nutrition; one can consume junk food with no nutritional value, or vitamin pills which have nutritional value but are not food. Through technologies like contraceptives and in vitro fertilization, even sex and procreation have become detached.

Much of the dis-integration and reintegration occurring on the technological plane is due to the flexibility of digital technology, which allows machine commands to be represented in a 'common tongue' and translated from one machine to another. Software applications are sold separately from hardware, and hardware is divided into modular components and accessories. Automation allows certain functions to be carried on without people, and away from their usual place of business; for example, automated teller machines are now found in many places where the bank has no branch office. And many kinds of transactions can now be carried out over the Internet as well.

The same flexibility that allows for the detachment of various components and functions means that they can be recombined in new ways, wherein a single machine carries out many different functions; for example, laptop computers can be equipped with cellular phones, e-mail retrievers, FAX machines, calculators, notebooks, etc. In one extreme example, computer consultant Steve Roberts built a computerized bicycle which, in addition to having 105 speeds, six brakes and pneumatic landing gear, is equipped with three computers, a cellular phone with modem and fax features, an answering machine, a CD stereo system, a ham radio, solar panels for power, a satellite dish receiver for sending or receiving electronic mail, and a security system that pages Roberts and dials 911 automatically so a speech synthesizer can report the theft.[3]

Through digital technology, reintegration can occur on scales previously not possible; worldwide computer networks, satellite communications, surveillance systems, electronically searchable archives containing millions of volumes —and these can allow for larger bureaucracies or government control. Networked systems with instant access capabilities and automated functions allow for data surveillance on a large scale. The US government's FinCEN (Financial Crimes Enforcement Network) is a database that can track financial transactions across the nation in real time, but can also be used for spying on people's accounts.[4] Automated systems which record transactions can be used to look into people's accounts, or even steal from them. The detachment of the teller from the bank has made a new kind of bank robbery possible, as reported in *Wired*;

> It could have been anywhere in the country - which is why the unprecedented deployment of a bogus ATM at a mall in Manchester, Connecticut, east of Hartford, sent a jolt through the hearts of bank-machine users everywhere. The fake ATM, brought in by brazen con artists who convinced mall officials they were genuine, recorded the card numbers and personal ID numbers of some 200 patrons, including a clerk from Kay's Jewelry, across from the ATM. . . . a machine that failed to deliver cash for more than a week was bound to elicit comment among the clerks at the stores around the mall court at Bucklands. Still, no one caught on. Then on Mother's Day two men arrived and wheeled the ATM away. The illicit cash withdrawals began soon after. The total mounted to US$50,000, then US$100,000 before they were caught.
>
> "This crime is the first of its kind in the world," says Dan Marchitello, the Secret Service agent who broke the case. The nearest precedent was a Staten Island case in which the culprits set up a videocamera [sic] with a telephoto lens and recorded patrons keying in their PINs, then correlated them with the times on discarded receipts.[5]

While the electronic surveillance of both video camera and database has become all too common, an industry is being built to mine the potential of telepresence, one of the most interesting examples of detachment and reintegration. Telepresence links a user to a robot at a remote worksite, where video cameras on the robot allow the user to see what the robot "sees", while the user remotely controls the robot's arms, position, and movement. The user can then carry out tasks at the worksite as if he or she were physically present there (the Mars Rover is one notable example). The degree of telepresence depends on the interface; some allow a high degree of control and interaction (for example, a completely mobile robot) while others are limited to a robot arm and a single video camera. Video conferencing in business is sometimes referred to as telepresence, even though the only connections are audio and video links.

Telepresence can be used in situations too dangerous for human beings; for example, in areas of high radiation, extreme temperature or pressure, underwater, or outer space. NASA's Dante project used telepresence to send a mobile robot down into a live Volcano in Alaska. Other uses include remote surgery, product testing and 'virtual showrooms', or art, education, and business. The robots can vary in size; miniature ones could be used to perform delicate internal surgery, or large ones could be used for construction or as military weaponry.

The idea of telepresence plays with the concepts of presence and absence which are at the heart of digital technology, reintegrating them; the user is sensorally *present*, but physically *absent*. In another sense, the term "telepresence" is an oxymoron; the user is clearly *not* present at the remote site, even though the robot *is*. The term, however, assumes the interface between the two to be so transparent as to be unimportant; indeed, transparency is one of the design goals in the design of telepresence technology. A person's actions are detached from their effects; the body is not held accountable for them. Normally, the results of touching fire or falling off a cliff will take their toll on the body, and so such behavior is limited accordingly. Interface transparency may help make the telepresent experience seem more immediate, but it cannot eliminate the distance between the user and the telepresent task, since the whole purpose of using telepresence is to do things that one normally could not.

The detachment of actions and consequences raises ethical questions concerning the use of telepresence, especially when it is used to interact with other human beings. Consider the following police arrest as reported by Gareth Branwyn in *Wired*;

> The "police officer" in this case is not one person but two human operators and a rather crude 480-pound robot dubbed the "Remote Mobile Investigator" (RMI). The arrest occurred on September 2,

> 1993, in Prince George's County, Maryland. Earlier in the day, Smith, 22, allegedly killed his girlfriend and raped another woman. After police had exhausted all other options for routing Smith from the building, they decided to borrow the local fire department's RMI-9 unit, a remote-controlled vehicle designed for handling hazardous waste and assisting firefighters. At a safe distance outside the apartment, a police captain and a robot tech used the RMI as their eyes, and finally as their fists, using a high-powered water cannon to clobber Smith.[6]

Later in the article, Branwyn quotes authors Hans Morevac and Manuel De Landa on their reaction to the incident;

> Does he [Morevac] have problems with the idea of cops making arrests from within such a virtual environment? "I think there are more problems with cops in the field than there will be with robots ... under these life and death situations. For one thing, the entire session will be recorded, so there will be greater accountability. Violence occurs through loss of control. Not having officers' lives in jeopardy will allow them to maintain their cool." Without reservation, Morevac concludes, "For them, it will be like playing a video game."
>
> This video game mentality strikes a raw nerve in Manuel De Landa. ... De Landa is concerned about the increasing use of AI and other advanced technologies designed to remove humans from the decision-making loop. ... De Landa is quick to shoot down the seemingly cut-and-dried "it's safer for our boys in the field" defense of a robocop future.
>
> "While I can see a good point in getting police out of harm's way, there is usually a political component to these arguments. The development and deployment of a new weapons system is rarely based solely on issues of safety and human concern," says De Landa. Asked how this specifically applies to a case against robocops, he replies, "It becomes a way of further distancing the cop from the suspect. It is difficult to hit or shoot another human being. It is easier if you have a teleoperated mechanical prosthesis doing it for you. There would be a desensitization here that I'd be concerned with."[7]

The connection between video games and war games has often been noted, as well as the desensitizing effects of both (for example, in films like *Wargames* (1983) and *Toys* (1993)). With sophisticated military technology and simulation, the gap between the two types of games is closing fast. Pilots and soldiers train in simulators before combat, and often rely heavily on machine mediation during combat, for weapons information and accuracy, visual information representing the environment and nearby enemy vehicles, and even decision-making. Battles can now be recorded digitally and are capable of being replayed and studied. Bruce Sterling writes about "The Reconstruction of the

Battle of 73 Easting", a tank battle which took place in February of 1991 in Kuwait during the Gulf War, and was later reconstructed in minute detail in an interactive computer simulation which can be viewed from any angle;

> "The Reconstruction of the Battle of 73 Easting" is an enormously interesting interactive multimedia creation. It is fast and exhilarating and full of weird beauty. But even its sleek, polygonal, bloodless virtuality is a terrifying thing to witness and to comprehend. It is intense and horrific violence at headlong speed, a savage event of grotesque explosive precision and terrible mechanized impacts. The flesh of real young men was there inside those flaming tank-shaped polygons, and that flesh was burning.
>
> That is what one knows - but it's not what one sees. What one really sees in "73 Easting" is something new and very strange: a complete and utter triumph of chilling, analytic, cybernetic rationality over chaotic, real-life, human desperation.
>
> Battles have always been unspeakable events, unknowable and mystical. Besides the names of the dead, what we get from past historical battles are confused anecdotes, maybe a snapshot or two, impressions pulled from a deadly maelstrom that by its very nature could not be documented accurately. But DARPA's "Battle of 73 Easting" shows that day is past indeed.[8]

The depiction of war through computer mediation or simulation is one that excludes the human factor, distancing viewers and participants from the human side of the drama.

Technology is now capable of physically removing soldiers from the battlefield as well, through the use of telepresence, remote control vehicles, and long-range weaponry like missiles and satellites. Barry Levinson's film, *Toys* (1993), plays upon this idea, tying it into video games as well. In the film, a military officer takes control of part of a toy company, to produce a miniaturized, remote control army. The miniature army will itself be controlled through a series of video game style interfaces, run by children trained on video games. The connection it makes between children playing warlike video games and computer-mediated remote control weaponry of the military is a literal one in the film, but it hints at the conceptual links that already exist. The desensitizing effects of warlike video games, coupled with the distancing effects of remote control and telepresence technology, help to bridge the gap between civilian and soldier, making the events of war seem gamelike and commonplace (not to mention the high levels of violence already found in movies and on television, which also act to desensitize viewers and acclimate them to a mediated environment).

By dis-integrating or detaching physical presence from the ability to perform actions that can have irreversible effects on a physical environment, actions become detached from their consequences. (Actions become detached both spatially, through telepresence and remote control, and temporally, through time bombs and other devices with delayed effects. As numerous experiments in psychology have shown, causes and effects are more likely to be associated with each other when they occur temporally and spatially adjacent.) The people responsible for such disembodied actions are not only removed from physical risk, but their absence at the site means that their knowledge of the consequences of their actions is limited to the narrow bandwidth allowed by the technology, and one which is often controlled by others. Even the empathy and sympathy of sensitive individuals can be lessened when consequences of an action are eliminated or filtered by others; a fact which the military may wish to exploit.

For every dis-integration due to technology, there are possible reintegrations, although usually technology-based ones; technology can propagate by attempting to solve the problems it creates. Whether the problems and solutions balance out, or in what direction the imbalance lies, are questions often at the heart of debates surrounding the appearance or implementation of new technology. Since one technology builds on another, solving its problems or bringing them to light, discussion can never be limited to a single technology; and as the numbers and kinds of technology grow, issues grow more complicated. Issues of dis-integration and reintegration can be intertwined and inseparable, but not necessarily balanced. Thus far most of my examples have focused on negative aspects of the technology, but there are cases in which digital technology has been used, not for dehumanizing effects, but to enhance identification with other human beings making their personhood less of an abstraction.

One such example is the sonogram, the use of ultrasound technology to create a moving image of a child in a mother's womb. Whereas discourse supporting abortion attempts to objectify the unborn child into a mere lump of tissue or even a non-living being, the sonograms produced with ultrasound (as well as the images produced with intrauterine photography[9]) help to affirm the personhood of the unborn child. In "Seeing the Baby: The Impact of Ultrasound Technology", Rita Beck Black writes;

> Previous studies about women's experiences during pregnancy and the development of maternal bonds to the baby have emphasized the perception of fetal movement as an important milestone (Bibring *et al.*, 1961). As the mother feels movement, she begins the process of knowing the baby as an autonomous individual. However, until

> recently, the pregnant woman was unable to "see" her baby. The mental image that she formed thus arose only from her imagination (Milne and Rich, 1981). As shown in the work of Lumley (1980), prior to the availability of ultrasound and other forms of prenatal technology, the majority of women only gradually came to see the fetus as a baby. Especially striking was the first trimester when most women only had a vague mental image of the fetus' shape and few described the fetus as a baby. Even well into the second trimester, Lumley (1980) found that only 44% of women described the fetus as a baby.
>
> Potential benefits of seeing the fetal image include reduction of maternal anxiety about the development of the fetus, resolution of early ambivalence about the pregnancy, and greater maternal attention to prenatal medical recommendations and healthful changes in eating and lifestyle (Fletcher and Evans, 1983; Reading and Cox, 1982). The introduction of ultrasound techniques during pregnancy has raised important questions about whether this technology accelerates the process of parental bonding (Fletcher and Evans, 1983). Pregnant women who have undergone ultrasound report "an increased sense of knowing their babies" (Milne and Rich, 1981, p. 32) and seem to feel closer to them (Kohn *et al.*, 1980). Seeing the baby move seems to further confirm its life and identity even if movement already has been felt prior to the ultrasound (Milne and Rich, 1981). The intensity with which women examine the ultrasound images of their fetuses seems not unlike the intensity with which women stare at their newborns in the early phases of bonding after birth (Klaus and Kennell, 1982).[10]

The enhanced reality provided by ultrasound helps to affirm the identity of the babies it images, but this identity can be affirmed without it as well; the mother, after all, may not be able to see the baby, but she can sense its presence in a variety of other ways. Feminist Ann Oakley suggests that one of the dangers of ultrasound is that objective, partial knowledge gained through the technology will take the place of patient-generated data, promoting medical over maternal expertise.[11] While it can be argued that ultrasound imaging adds to a mother's knowledge rather than replaces it, the intervention of technology in natural processes and experiences cannot be ignored.

Digital technology's *dis*-integration and *re*integration, of both the environment and our experience of it, continue as technology is implemented in the natural world, and in the processes that go on there. While its effects can be positive as well as negative, it still, in the end, increases the level of abstraction present in human experience, and carries forward the engulfment of the individual in mediation.

Approach and Engulfment

In his famous essay "The Myth of Total Cinema", Andre Bazin describes how the early pioneers of cinema "saw the cinema as a total and complete representation of reality; they saw in a trice the reconstruction of a perfect illusion of the outside world in sound, color, and relief."[12] Bazin sees their efforts unified towards a common goal;

> The guiding myth, then, inspiring the invention of cinema, is the accomplishment of that which dominated in a more or less vague fashion all the techniques of the mechanical reproduction of reality in the nineteenth century, from photography to the phonograph, namely an integral realism, a recreation of the world in its own image, an image unburdened by the freedom of interpretation of the artist or the irreversibility of time. If cinema in its cradle lacked all the attributes of cinema to come, it was with reluctance and because its fairy guardians were unable to provide them however much they would have liked to.[13]

Today, new media like virtual reality and telepresence are extending the myth; through digital technology in the form of the stored program computer, interactivity has been added to enhance realism. Indeed, according to Bazin's myth, cinema has always been striving to become "virtual" reality.

To be more precise, the "Myth of Total Cinema" applies mainly to form, perceptual realism, and experience, than it does to content. Although Bazin uses the phrase "reproduction of reality", the term *representation* comes closer to the truth, especially when its intense partiality is acknowledged, which Bazin had hoped could be minimized. According to the myth, it would seem that cinema's teleological goal (or rather, the goal of its inventors) is the replacement of consciousness, or at least control of all of a viewer's sensory input. Nor need this goal be limited to cinema alone, as it applies to media in general, which replace direct experience with mediated experience.

If we personify the forces directing the development of media technology, we can create a metaphor similar to those used by technological determinists, who see technology as a force which works at achieving a goal; while such a metaphor is of course a simplification, it does provide an interesting way of looking at the development of media technology. Through this metaphorical personification, we can describe a "Myth of Total Media" in which different forms of media conspire together to allow the viewer to lead an increasingly mediated existence. Digital technology, through multimedia, allows greater cooperation between media, providing a common tongue by which they can work more closely together. Through various stages over the last hundred years or so, these media have been physically approaching closer and closer to their audience, and gradually engulfing them, enfolding their senses in a digital environment. To see how this strategy of approach and engulfment has unfolded, we can look at various forms of mediated experience, particularly in terms of distance, audience, and sensory engulfment.

For centuries since the early Greek and Roman times, there was the theater, with its stage which became a different space during a performance; it could be any time or place imagined by the playwright. The stage was set off from the audience, and often was raised off the ground to provide a better view to patrons sitting further away (some of the ancient amphitheaters could seat thousands). Most patrons were a good distance away from the action onstage, so makeup, gesture, and speech were forced to compensate, amounting to a style of performance we now refer to as "stagy". While the original Greek and Roman amphitheaters were open air, later designs were built within walls, and later roofed over, during the 17th century. As the theater moved indoors, lighting and sound could be better controlled by stage hands, but the audience became cut off from nature; their experience could now be more closely controlled. The lighting in the theater could be dimmed, and the stage lit up; encouraging audience members to look at the stage instead of each other. Indoor stages meant smaller audiences, and ones closer to the stage, the closer proximity enhancing their experience.

When cinema appeared in the 1890s, it had theater as a model of exhibition. But whereas the stage had physical depth, film brought everything closer to the audience, in one sense, flattening the depth of the stage onto a screen, just as the painting and photograph had done.[14] Unlike theater before it, the cinema *had* to be darkened, in order for the projected image to be seen clearly. Unable to provide feedback to the actors onscreen, the audience became more physically passive; and while films were more mediated than theater, they were not limited to the stage, and could take their viewers anywhere.

The audience was also "brought closer" in an experiential sense, through medium shots and close-ups. The theater did not engulf one the way the cinema did; the cinema constructed a point of view for the spectator, addressing its audience on an individual basis, unlike live theater, where each person's vantage point was slightly different, the actors having to play to them all. Cinema could create points of view in ways that theater could not, drawing the spectator "into" the film.

The cinema was more intimate, visually, and filmmakers could convey subtleties through the close-up far greater than what any stage director ever could do on-stage. The cinema could also potentially take up more of an audience member's time; although films were initially shorter than plays, there were far more of them available, at more times during the day; they changed more often than plays; and they were cheaper to attend.

In the twentieth century, the development of radio, and then television, brought media even closer to the viewer. Physically, there were only a few feet between the cabinet and the listener or viewer; the television screen was far smaller than the cinematic screen, but its proximity helped to compensate. It was also physically closer to the viewer in the sense that it was in the home; whereas one had to "go out" to the movies, the radio and television were always available and convenient; they could be owned by the viewer, and their programs were even cheaper than the movies —free, once the sets were bought. The number of daily hours of radio and television programming increased over time and there were far more radio and television programs available than there were movies, since theaters showed the same films over a few times a day for weeks on end. Television shows were shorter and changed faster than movies, and people could easily spend several hours a day watching. Today, of course, there are more radio and television channels than ever, many of them on the air twenty-four hours a day. Even cable channels that have monthly charges are still far cheaper, hours per dollar, than theatrical cinema could ever be. Movies, as it has been well documented, tried to win audiences back to the theater with, among other things, widescreen and 3-D, two strategies designed to bring the image "closer" to the audience. But movies could not provide entertainment as continuous or as cheaply and conveniently as television could.

By moving into the home, the viewing audience was broken up into small segments or individual viewers (the darkness of the theater, which had previously fulfilled that task, was no longer needed). The stage and movie screen had both been objects lit from afar; but the television was itself a source of light, as were all the cathode ray tube-based media that came after it. The lack of audience did create a few problems; would audience members at home know when to laugh or react, and might they miss the collective experience of film and theater? But these were solved with devices like canned laughter and the "live studio audience" with whom you were watching the show, who could be heard —and often seen— to assure you that you weren't watching alone. Later advances along these lines included radio and television call-in programs, where the opportunity of feedback existed, even though a very limited number of calls could actually be taken. Whereas cinema's construction of spectator point-of-view helped individualize the audience, television tried to help its divided viewers reestablish the feeling that they were still a part of the audience crowd.

Video cassette recorders gave viewers some control over television programming, and made shows even more available than they were on the air; you could see them any time you wanted once they were on tape. The VCR increased the availability of movies as well, just as the audio cassette tape, 8-track tape, and compact disc made music available independent of radio scheduling or live concerts. After the appearance of the radio, sound technology approached closer to the listener as well; speakers grew larger in size as high-fidelity equipment appeared, and radios shrunk from furniture-sized receivers to countertop models, pocket transistor radios, and finally radios worn on the body itself. External speakers gave way to headphones worn on the listener's body, addressing the listener individually while blocking out external sound. Headphones covering the ear shrunk to earphones laid on the ear, to mini-earphones placed *into* the ear. Media technology was not only physically closer to the body, it was entering it.

Visual media continued to approach the viewer ever more closely as well. Whereas VCRs brought the viewer closer and allowed very limited interaction (the changing of tapes), the video game let the viewer become a player, with interaction that was onscreen and instantaneous. Before it, too, entered the home, the video game made its first appearance as an arcade game, in the tradition of the pinball game, whose mode of exhibition it shared. Before long, a version of *Pong* was available for the home, along with dozens of other home video game systems. The video game brought the television even closer to the viewer; in the arcade one stood right in front of it, inches away instead of feet. For many families, video game systems were the first computers in the homes; and as the video game market opened up in the late 1970s and early 1980s, so did the home computer market.[15] The computer brought the CRT screen even closer to the user, right up to the desktop, and allowed even more interaction than video games had (many computer programs were games as well, but other uses, such as word processing or spreadsheets, were also available). The paddle, joystick, trak-ball, keyboard, and mouse brought the user into physical contact with the medium, and pocket-sized video games run on batteries could be carried upon one's person, making play available anytime, anywhere.

Although many video games are two-player, most of them have one-player modes where the computer supplies the other player (just as canned laughter simulated the rest of the audience, the computer can simulate other players). Most applications of the computer, however, are not games at all; word processing, data entry and retrieval, design, simulation, imaging, and so on, are its main applications in industry. While some uses like e-mail and conferencing involve other people, much computer use is a solitary activity. The computer thus is able to take up more of the average person's time by extending beyond entertainment and informational uses; instead of remaining purely a free-time activity, it is used throughout the workday by an increasing number of people. It is not merely a source of fantasy, but a window through which reality is seen and envisioned.

Still largely in development, virtual reality brings the screen closer still, within a few inches of a person's eye. One's entire visual field is taken up by the head-mounted display, while datagloves cover users' hands. Sensory engulfment is acknowledged with the term "immersion" which is applied to a number of technologies, including virtual reality, telepresence, and simulator rides which enclose the user's whole body. These digital developments are ideas with long histories; virtual reality's bringing of the image right up to the eye dates back to stereo viewers, the mutoscope and kinetoscope, and 20th century optical toys like Mattel's Viewmaster. Immersive technology's ancestry can be traced back to the camera obscura, and through later developments such as the panoramas and dioramas of the late nineteenth century, through military flight simulators, and mechanical theme park rides.

Although some simulator rides, like Lucasfilm's *Star Tours* ride at Disneyland, are miniature theaters holding 20 or so people, most interactive immersion technology is made for individual users. The partial immersion of "augmented reality", discussed in the next chapter, will also find more uses and may even become more widespread than virtual reality. Newer and developing immersive technologies draw even closer to the individual; Tom Furness at the University of Washington's Human Interface Technology (HIT) Lab in Seattle has built display devices that use a low-powered laser to print images directly on the retina. In this case, the technology is not only worn on the user's body, it creates its images *inside* the body, focusing them within the eye.

The next step, of course, is to bypass the eye entirely and go directly to the brain itself; the final stage of sensory engulfment. Just as digital encoding into bits provides a common form of expression for various types of data, the electronic impulses of the brain's neurons are used by it to represent all the senses. The brain is the final destination for the approach and engulfment of digital technology, which through brain implants would allow it to control sensory experience directly. Precursors for this sort of control are drugs, such as LSD, and chemicals that enter the brain through the bloodstream and work directly on the physical matter of which it is made. Their uncontrollable effects, however, render drugs unsafe; once they enter the brain, they may reside there for some time and alter brain chemistry. Nor can the experience be turned on and off at will; digital technology is expected, by some, to provide forms of neural stimulation control not possible with drugs.

While serious researchers are assessing the possibilities of using implants to provide vision for the blind and have already had some success in improving hearing for the deaf,[16] others are interested in the idea of brain implants for purely recreational purposes. There are a few people conducting research in this area, like composer David Rosenboom, who is working on a direct interface to the music composer's brain,[17] but most cyber-enthusiasts are looking for pleasure or power. William Gibson's novel *Neuromancer* features characters "jacking" themselves into "simstim", short for simulated stimulation, by plugging wedge-shaped chips into an interface port implanted in their skull, directly accessing their brain. While neural implants are still mainly the product of science fiction, the idea has attracted hobbyists who are experimenting at home. In a subsection entitled "Basement Neurohackers" of his article, "The Desire to Be Wired", Gareth Branwyn writes;

> Perhaps more in the realm of science fiction than science fact, "neurohackers" are the new do-it-yourself brain tinkerers who have taken matters into their own heads. "There is quite an underground of neurohackers beaming just about every type of field imaginable into their heads to stimulate certain neurological structures (usually the pleasure centers!)," a neurohacker wrote to me via e-mail. Several of those basement experimenters were willing to talk. . . . David Cole of the nonprofit group AquaThought is another independent researcher willing to explore the inside of his own cranium. Over the years he's been working on several schemes to transfer EEG patterns from one person's brain to another. The patterns of recorded brain waves from the source subject are multiplied many thousands and then transferred to a target subject (in this case, Cole himself). The first tests on this device, dubbed the Montage Amplifier, were done using conventional EEG electrodes placed on the scalp. The lab notes from one of the first sessions with the Amplifier report that the target (Cole) experienced visual effects, including a "hot spot" in the very location where the source subject's eyes were being illuminated with a flashlight. Cole experienced a general state of "nervousness, alarm, agitation, and flushed face" during the procedure. The results of these initial experiments made Cole skittish about attempting others using electrical simulation. He has since done several sessions using deep

> magnetic stimulation via mounted solenoids built from conventional iron nails wrapped with 22-gauge wire. "The results are not as dramatic, but are consistent enough to warrant more study," he says.
> Part of the danger of monkeying with one's brain, especially with little or no knowledge of neuroscience, is that most individuals do not have access to the sophisticated testing and feedback devices that are available to legitimate researchers. ... The further I got out on the fringes of neurohacking, the more noise overcame signal. I heard rumors of brain-power amplification devices, wire-heading (recreational shock therapy), and most disturbing of all, claims that people are actually poking holes in their heads and directly stimulating their brains.[18]

Should brain implants and recorded brain waves become a new source of mediated experience, digital technology will have completed its quest for sensory engulfment and total immersion; it will have literally become integrated directly into the user.

Such speculation aside, media technology has been given a large role in many people's daily lives, and mediated experiences abound. When one considers how much what most people know of the world is known not through first hand experience but rather through mediation, the difference is staggering. Although the above paragraphs personify media technology as a force with a teleological goal, it is, of course, human beings who choose to implement and use the technology. Through media we are able to get an overall, global image of things and events that would be very difficult to attain through direct experience; but the price paid is the reliance on the technology of mediation (and those in control of it). The time and attention spent on the global view is time and attention taken away from the local view and direct experience; and since the average individual has more influence over local events than global ones, this in turn may cause people to feel more ineffectual and distanced from events.

The mediated and reconstructed world the media offers its viewers, in exchange for their limited time and attention, and the increasing interactive possibilities and technologies which claim to empower the individual, are saying, in effect, that "the world can be under your control (or at least gaze) if you allow technology to mediate it for you". The image of media technology extending the world to individuals in return for their time and attention —and belief in it— sounds eerily similar to the Temptation in the Desert;

> Then the devil took him up higher and showed him all the kingdoms of the world in a single instant. He said to him, "I will give you all this power and the glory of these kingdoms; the power has been given to

> me and I give it to whomever I wish. Prostrate yourself in homage before me, and it shall be yours."[19]

Media technology, however, tends not to be as overt in its tempting; when it shows us the world in an instant and promises power at a distance, tacit acceptance of its claims is already complicity. Its usefulness and omnipresence make acceptance seem natural; and the more omnipresent it grows in society, the more it is taken for granted, until it is simply part of the background environment.

Abstraction and Invisibility

We live in an age in which practically anything can come to depend on electricity; there are electric toothbrushes, electric car windows, and even electronic books. Electrical wires run through walls, ceilings and floors, underground and overhead, but go largely unnoticed. Technology is often relied upon when people eat, travel, and communicate; for many, it is as ubiquitous and depended upon as the air they breathe. But this was not always the case; for centuries —or even millennia— machines were far more noticeable, and stood out from the otherwise natural environment surrounding them.

Technology can fade into the environment only after its presence has been effaced, and ironically, this effacement relies on media and public attention. The attention and enthusiasm given new technologies help them become widespread, and only after they are widespread can they become commonplace, and ready to fall from public notice as newer, more unusual technologies steal the spotlight. In addition to its familiarity, technology's growing complexity has helped abstract it and efface its functioning, making it much easier to take for granted.

For a long time in history, all machines were large, physical objects, with visible moving parts; gears, levers, belts, pulleys, weights, and wheels were set in motion when the machine was at work, and one could follow the chain of cause and effect throughout its workings. After watching it awhile, one could obtain some sense of how the thing worked; and someone familiar enough with it might be able to determine what was wrong with it when it didn't work, could even repair it. As clocks developed, their gear chains grew more complex; gears were stacked on axles, escapements and springs were added, and a gearbox had to hold all the pieces in place without hindering their movement. Intricate clockwork was more delicate than other simpler devices, and its performance could be better guaranteed if it was shielded from the elements; so its workings were enclosed in a box. Some windmills had their grinding gears enclosed inside a tower, but unlike the clock, one could walk inside a windmill to observe its functioning. The enclosing of the machine in a box meant its workings were no longer visible, making it more difficult to understand (the increasing complexity of gear systems also helping to obscure the function of any particular part). Technology was beginning to fall from visibility.

Smaller and smaller clocks were built, and the more portable they became, the more widely they were used. Carriage clocks were the first clocks small enough to be portable, and later wristwatches appeared; miniaturization was the next step in decreasing visibility. Even if the workings could be seen, they were difficult to discern; and the casing was more important than ever to protect smooth functioning. The machines' greater complexity meant experts were needed to make repairs and adjustments, and even to determine exactly what was wrong in the first place.

The rise of magnetic and electrical technology in the late nineteenth and early twentieth century took the notion of invisibility another step further. Unlike the mechanical devices that came before them, electrical devices had fewer moving parts (or none), and were powered by an invisible force "flowing" through the wires.[20] One could not see the movement of electricity in the wires; all that remained visible of the machine's functioning were the mechanical machine parts controlled by electricity. And in many devices, such as radios, there were very few or no moving parts. Electrical signals allowed messages to become detached from the media they appeared in; the message could be sent without any physical object being sent. With the telegraph, the message became invisible as well. In previous media, such as writing, painting, photography, sheet music, pinned barrels, perforated discs, or punched cards, messages were made of visible differences or variations in the material itself; paint, chemical deposits, marks made on paper, or holes or pins. But variations of electric current making up a signal, or magnetic particles on magnetic tape, are too small to be seen (although one can hear an audio signal of data transmission through a modem, the signal is far too fast to hear individual bits).

Next came the expansion of electric technology into electronic technology, which made the stored program computer possible. The difference between electric and electronic technology is perhaps best exemplified in the difference between the electric and the electronic typewriter. The electric typewriter differed from the manual typewriter in that an electrical motor provided power for all functions initiated by hitting the keys, while the electronic typewriter used a microprocessor and had a "memory", allowing it to do word processing, in which a series of words (or lines or paragraphs) could be stored in memory and manipulated before being printed out. Thus it enabled the process of *typing* to be detached from that of *printing* the characters out on paper. Microprocessors and computer technology, which were electronic, eliminated more of the moving parts found in nonelectronic electrical technology, and carried the miniaturization process further, as technology moved from the vacuum tube to the transistor, to the integrated circuit, and to smaller and smaller printed circuits. In integrated circuitry, there are no moving parts, although the program is said to be "running" within it.

The idea of software has changed the definition of machine. David Gelernter refers to computer programs as "disembodied information machines", for which the computer provides a body.[21] Computer software, seen as a machine, is completely made of electronic impulses and invisible. The code of a program can be printed out, but the printout it is not the *machine*, only a representation of it. For software to function as a machine and perform a task, it can only exist in electronic form.

Physical machinery as well is still shrinking in size, towards invisibility. Nanotechnology promises virus-sized molecular machines that can replicate themselves, and unimaginably tiny machinery and computers. As K. Eric Drexler writes;

> Brass gears and Tinkertoys make for big, slow computers. With components a few atoms wide, though, a simple mechanical computer would fit within 1/100 of a cubic micron, many billions of times more compact than today's so-called microelectronics. Even with a billion bytes of storage, a nanomechanical computer could fit in a box a micron wide, about the size of a bacterium. And it would be fast. Although mechanical signals move about 100,000 times slower than the electronic signals in today's machines, they will need to travel only 1/1,000,000 as far, and thus will face less delay. So a mere mechanical computer will work faster than the electronic whirlwinds of today.
>
> *Electronic* nanocomputers will likely be thousands of times faster than electronic microcomputers—perhaps hundreds of thousands of times faster...[22]

The possibilities of nanotechnology are said to include machines that can make copies of themselves, much as viruses do, only they could be made from inorganic matter. As programmable machines, they could be set to reproduce in great numbers, and rebuild the physical environment. Drexler acknowledges the potential harm that they could bring about;

> The early transistorized computers soon beat the most advanced vacuum-tube computers because they were based on superior devices.

> For the same reason, early assembler-based replicators could beat the most advanced modern organisms. "Plants" with "leaves" no more efficient than today's solar cells could out-compete real plants, crowding the biosphere with an inedible foliage. Tough, omnivorous "bacteria" could out-compete real bacteria: they could spread like blowing pollen, replicate swiftly, and reduce the biosphere to dust in a matter of days.[23]

Drexler also describes how nanomachines could be used inside the human body to repair cells and tissues. While nanotechnology does not yet completely exist, microelectronics have already shrunk beyond visibility and are even beginning to merge with biotechnology.[24]

Machines have so greatly changed in definition as well as physical form, that the public's conceptualizations of them have had a hard time keeping up. As abstraction and invisibility make an understanding of a machine's functioning more difficult and increasingly limited to experts, several strategies have developed to help bring a limited understanding of them to the public, whose support is needed if the machines are to come into common use. Once a sufficient level of public understanding is reached, a technology can become accepted, widespread, and eventually taken for granted, allowing it to retreat into the background environment, paving the way for further advances building upon it.

Typical strategies for the public understanding of a complex new technology include the organic metaphor, simplification and comparison, and dismissal of the complexity. The organic metaphor is quite common; in Drexler's quote above, for example, nanomachines are compared to bacteria and plants with leaves. The growth of biology as a science helped to make biological processes seem mechanical, as cause and effect relationships were established between the various "parts" into which living organisms were divided. Viruses and other such discoveries questioned the boundaries between living and nonliving, and organic and inorganic. As science grew, biological and mechanical concepts came to resemble each other and metaphors were drawn between them (after all, as it was often pointed out, there were electrical impulses running in both). The organic metaphor remains an easy way to describe a complex technology to the public, because it focuses on what the technology *does*, rather than how it *works*. While the inner workings are usually far from those of biological organisms, such metaphors work by giving the machine a framework of useful, predictable behavior in the eye of the public. People are more accepting of the technology as long as they know what to expect from it.

Another way to shape expectations is to simplify the technology and compare it to an earlier, well-understood one that is already commonplace. Despite great differences between new and old technologies, and novelty and new uses as important marketing tools, the simplification of complex technology for a nontechnical public seems inevitable. Automobiles were "horseless carriages", and film was seen as "moving pictures" and later "talking pictures". Television was radio with pictures, and even contained many of the same shows and personalities which made the transition from one medium to the other. Word processors were first seen as typewriters with memories, and computer keyboards still uses the "QWERTY" key arrangement of the typewriter. Simplification, especially of highly complex technologies like the computer, often begins to border on metaphor (for example, the icon-driven graphic user interface), and is also usually more focused on *what* something does instead of *how* it does it.

A third strategy is that of dismissing the complexity, which acknowledges the change of focus and waves off understanding of the technology as unimportant, and not needed for its use. The most common expression of this sentiment is the cliché that "you can drive a car without knowing how the engine works". What it fails to mention is that you can only drive a car *if* it works, and you have to know how it works to fix and maintain it, if it is to work properly. Often, however, a knowledge of what something does is enough for most users, who may have enough trouble learning how to use it, much less understand its inner workings.

Metaphors and comparisons to existing technologies are also often apparent in the physical design of the technologies, to help blend them into the background of the environments where they appear. The radio, when it first appeared, came in a large wooden cabinet, and fit into the home by masquerading as a piece of furniture. The wooden cabinet was also used by televisions into the 1970s, and even the Atari 2600 home video game system had a plastic wood-grain panel in front. Likewise, computer icons like the paintbrush, trash can, pencil, eraser, scissors, and so on indicate computer processes that are functional metaphors of physical ones.

One need only consider how quickly computer technology has become commonplace in everyday life to see how the flexibility and connectivity possible with digital technology makes it particularly easy to blend into existing technologies and their environment. Whether the environment it creates is a physical one or an experiential one, digital technology has become woven into daily life, an interface to the world for many. Once it is worked into one's surroundings, ignored and taken for granted, digital technology changes our existence in the environment; it changes what we think of as reality.

NOTES

1. Ellul, Jacques, *The Technological Society*, translated by John Wilkinson, Foreword by Robert K. Merton, New York: Alfred A. Knopf, Inc., ©1967; on the opposition to liberalism, see pages 200-205; on the opposition to democracy, see pages 208-218. For the source of the quote, see page 21.
2. Ibid., pages 137-138.
3. See Cassidy, Mike, "Ultimate bicycle took shape in Mountain View", *San Jose Mercury News*, May 23, 1993, page 1B(2), and Dern, Daniel P., "Easy rider: a restless computer consultant takes his business and office on the road", *Home Office Computing*, Volume 10, Number 10, October 1992, page 62(2).
4. Kimery, Anthony L., "Big Brother Wants to Look Into Your Bank Account", *Wired*, December 1993, page 91. Even on-line services can have the capabilities of scanning your hard drive; one program was used to scan customers' computers to determine the type of microprocessor, the version of DOS and Windows, the type of display and mouse, and the amount of free space available on the hard drive. See "Online Spying", *Forbes*, February 13, 1995, page 186.
5. Patton, Phil, "The Buckland Boys and Other Tales of the ATM", *Wired*, November 1993, pages 46 and 48.
6. Branwyn, Gareth, "Hard-Nosed Cops? Crime in the age of intelligent machines", *Wired*, April 1994, page 62.
7. Ibid., page 65. Ironically, the page facing the last page of Branwyn's article is a full page advertisement for "Comanche CD", a helicopter war simulation CD-ROM, which brags about how real the game is.
8. Sterling, Bruce, "War is Virtual Hell", *Wired*, Premier Issue 1993, pages 95-96.
9. For examples of intrauterine photography see Voyager's *Miracle of Life* laserdisc, and the work of photographer Lennart Nilsson, including his book, *A Child is Born: The drama of life before birth in unprecedented photographs*, photography by Lennart Nilsson, text by Axel Ingelman-Sundberg and Claes Wirsen, translated by Britt and Claes Wirsen and Annabelle MacMillan, New York: Delacorte Press (1st American edition), ©1966, 1967. See also Bentley, Joelle, "Photographing the miracle of life: the work of Lennart Nilsson", *Technology Review*, Volume 95, Number 8, November/December 1992, page 58(8).
10. Black, Rita Beck, "Seeing the Baby: The Impact of Ultrasound Technology", *Journal of Genetic Counseling*, Volume 1, Number 1, 1992, page 46.
11. See Oakley, Ann, "The History of Ultrasonography in Obstetrics", *Birth: Issues in Perinatal Care & Education*, March 1986, Volume 13, Number 1, pages 8-13.

12. Bazin, Andre, "The Myth of Total Cinema", in *What is Cinema? Volume 1*, essays selected and translated by Hugh Gray, Berkeley and Los Angeles, California: University of California Press, ©1967, page 20.
13. Ibid., page 21.
14. Ironically, cinematic precursors like the mutoscope, the kinetoscope, and the kaleidoscope, were ahead of their time in terms of their close physical proximity to the viewer, since they came right up to the viewer's eyes and demanded the full field of vision. But their flipping cards and extremely short programs could not provide the same sensory richness as projected films, which could also run for a longer time. Thus the projected form of cinema won out, and the displays brought up to the eyes had to wait for the Viewmaster and especially virtual reality to once again take hold of an audience.
15. Theater's influence on the design of new media continues even in the computer age; according to an interview with John Kador, Alan Kay, the designer who thought up object-oriented programming, graphical icons, windows and the Dynabook, invented the Macintosh graphical user interface (GUI) by thinking of computers as theater. See Kador, John, "One on one", *MIDRANGE Systems*, Volume 6, Number 4, February 23, 1993, page 38.
16. For an overview, see Branwyn, Gareth, "The Desire to Be Wired", *Wired*, September/October 1993, page 62. For articles on artificial vision and hearing, see Leutwyler, Kristin, "Prosthetic Vision; Workers resume the quest for a seeing-eye device", *Scientific American*, March 1994, page 108; Miller, Jonathan, "Digital implant awakens sound in deaf ears", *New Scientist*, February 11, 1989, page 40; Dettmer, Roger, "The all-electric ear", *IEE Review*, May 1988, pages 195-198; Erickson, Deborah, "Electronic Earful; Cochlear implants sound better all the time", *Scientific American*, November 1990, page 132; and Pfingst, Bryan E., David R. De Haan and Lisa A. Holloway, "Stimulus features affecting psychophysical detection thresholds for electrical stimulation of the cochlea", *The Journal of the Acoustical Society of America*, October 1991, pages 1857-66, among others.
17. As reported by Roger F. Malina in "Digital Image—Digital Cinema: The Work of Art in the Age of Post-Mechanical Reproduction", *Leonardo*, Digital Image—Digital Cinema Supplemental Issue, 1990, page 36.
18. Branwyn, Gareth, "The Desire to Be Wired", *Wired*, September/October 1993, pages 65 and 113.
19. Luke, Chapter 4, Verses 5-7, translation from the *New American Bible*.
20. In one sense, we could say that the "moving parts" were the electrons themselves, taking the idea of smaller and smaller moving parts to an extreme. Occasionally complaints arise among users of technology which has too little visible; see Duntemann, Jeff, "The tragedy of the black box. (Motorola's MC3362 integrated circuit for ham radios as a metaphor for what is wrong with Turbo Vision: users cannot see what is going on inside

the system)", *Dr. Dobb's Journal*, Volume 16, Number 12, December 1991, page 123(5).
21. Gelernter, David, *Mirrorworlds: or the Day Software Puts the Universe in a Shoebox... How it Will Happen and What it Will Mean*, New York and Oxford: Oxford University Press, ©1991, page 39.
22. Drexler, K. Eric, *Engines of Creation: The Coming Era of Nanotechnology*, New York: Anchor Press, Bantam Doubleday Dell Publishing Group, Inc., ©1986, page 19.
23. Ibid., page 172.
24. Stix, Gary, "Gene readers; microelectronics has begun to merge with biotechnology", *Scientific American*, January 1994, page 149-50.

9.
Virtual Reality and Other Substitutes

The virtual reality technology commercially available today is still relatively crude compared to what the human sensorium is capable of experiencing, and more developed experimental strands of virtual reality are accessible only to a privileged few. Yet despite its physical absence in the public domain, virtual reality (VR) is present in many different forms in discourse throughout American culture and has had an impact on the public imagination. Whereas existing digital technology has physically changed the environment in which it appears, the technological dreams and fantasies surrounding virtual reality have changed what people see as the *potential* of new technologies. Film and television in particular have latched onto the technological fantasies surrounding new media, and even those beyond what is considered possible. Thus, virtual reality —or rather, the *idea* of virtual reality, in its various incarnations— has become an influential technology before becoming widespread, and its absence has encouraged speculation about its uses and capabilities.

This chapter will not be a history of virtual reality, nor a catalog of various kinds of VR, but rather a look at its place in culture and the way it relates to our experience of everyday life. VR fantasies almost always involve mediated reality and its resemblance to direct, physical experience; not surprising, perhaps, considering that as virtual reality grows more realistic, the physical world grows more mediated. Exploring this relationship, we shall first examine definitions and differences, and how terminology used by the industry reflects or effaces those differences. We can then looks at VR's place in fiction, especially in film and particularly in the *Star Trek* television series, and look at VR as entertainment. And finally, we will look at how technology, used as a substitute, is supplanting various experiences and making our conceptualization of the world more perceptual and more virtual than material. As Michael Benedikt notes,

> On the largest view, the advent of cyberspace is apt to be seen in two ways, each of which can be regretted or welcomed; either as a new stage in the *etherealization* of the world we live in, the real world of people and things and places, or, conversely, as a new stage in the *concretization* of the world we dream and think in, the world of abstraction, memory, and knowledge.[1]

"Virtual Reality", then, may be thought of as referring not to the product of a technological industry, but rather a philosophical condition brought on by mediation and the desire for fantasy and escape from the constraints of the material world around us, a state very different from that of direct experience.

Reality vs. Virtual Reality

Before we can define "virtual reality" we must define "reality", which has always been a philosophically slippery term. Discourse surrounding virtual reality and other mediated experience has brought about a return of the "real", providing a much-needed standard by which to judge mediated experience; an idea we find in terms like "*virtual* reality" or "*augmented* reality". Michael Heim sums up the history of the term "reality" in the Western world;

> Plato holds out ideal forms as the "really real" while he denigrates the raw physical forces studied by his Greek predecessors. Aristotle soon demotes Plato's ideas to a secondary reality, to the flimsy shapes we abstract from the really real—which, for Aristotle, are the individual substances we touch and feel around us. In the medieval period, real things are those that shimmer with symbolic significance. The biblical-religious symbols add superreal messages to realities, giving them permanence and meaning, while the merely material aspects of things are less real, merely terrestrial, defective rubbish. In the Renaissance, things counted as real that could be counted and observed repeatedly by the senses. The human mind infers a solid material substrate underlying sense data but the substrate proves less real because it is less quantifiable and less observable. Finally, the modern period attributed reality to atomic matter that has internal dynamics or energy, but soon the reality question was doomed by the analytical drive of the sciences toward complexity and by plurality of artistic styles.[2]

For our purposes here, then, it is perhaps more useful to ask not "What is reality?", but rather, "What do we mean when we say 'reality'?"; we need only say enough about it to make it distinct from the other states with which it is compared. For a discussion of mediation, we might define "reality" as the state in which the *least possible amount of mediation* is occurring. One could argue, perhaps, that there cannot be a state in which no mediation is occurring, since the eye, ear and other sense organs mediate between the brain and the outside world, and cultural codes and experience limit and shape perception. Since this mediation will occur in any case, we must look for the state of the *least possible* amount of mediation, in order to have some starting point that we can refer to as "unmediated" (and likewise so that "mediated" will have a specific meaning). We could say that "mediated" refers to *instrumental* (as opposed to cultural) mediation having origins which are *external* to the physical individual (and thus includes all devices, such as brain implants, which originate outside the individual before being implanted). While *instrumental* forms of mediation can be removed from a subject, *cultural* forms of mediation (such as language) cannot be stripped away, as they play a role in the constitution of the individual. Thus it is the instrumental forms, which can be removed, that will be discussed here.

By this definition, "unmediated reality", or "reality" for short, would refer to our direct, lived experience, the consensus of the senses, each verifying the data provided by the others and used to construct a detailed, robust, mental image of the world around us. One could then ask a question that appears often in science fiction; when confronted with conflicting information or two states each claiming to be reality, how do we know which of them is more real (or less mediated)? In most cases, we make our best guess, seeing how many of our senses can verify the data. Is the experience repeatable and consistent? Many people, on seeing a hologram, for example, undergo a similar process; they move around to view it from different angles (checking for consistency of form) and wave their hand through it as if to physically verify that the space it appears to occupy is really empty.

This, then, is the promise and allure of virtual reality; it is a mediated experience that provides a greater consensus of the senses than previous mediated experiences. As Gene Youngblood points out, "The world in the mirror is the archetypal virtual space".[3] The world in the mirror, however, is little more than one behind a window, and is not an interactive one (Lewis Carroll's *Through the Looking Glass* excepted). Dreaming and memory can provide integrated multisensory experiences that seem real, but they are wholly internal and noninstrumental. Virtual reality integrates sight, sound, and touch (in the case of force-feedback datagloves), and does so in real time (or at least attempts to, since in most systems, a slight lag remains between actions and on-screen actions). The machine's reactions to our actions are made to appear *as* our actions. As lag time decreases and resolution increases, VR mediation will be more transparent and the illusion of directness will improve.

Virtual reality technology, however, will always have certain limitations. To begin with, virtual reality itself involves a physical experience of some kind; for example, the wearing of a head-mounted display (HMD) and dataglove, which often must be calibrated before use, or the enclosing of the body within a simulator. Virtual reality experiences, are, in this sense, a subset of real world experiences, although these physical aspects of the VR experience are usually effaced. The weight, size, and design of the interface will no doubt change to decrease the user's physical awareness of the interface, but the technology will always be present in some form. Despite its limitations, however, VR technology can already present an illusion strong enough to elicit physiological and psychological effects.

Just as some people feel motion sickness during an IMAX film, the loss of a frame of reference during a VR experience can result in "simulator sickness", believed to be caused by conflicting sensory input. After spending even a short time in an immersive VR experience, users have experienced disorientation, dislocation, sweating and exaggerated reflex responses, dizziness, upset stomach, headaches, eyestrain, and lightheadedness, and severe nausea. In one study with 112 subjects, 40% of them reported symptoms of dysequilibrium, including mental symptoms of confusion, depression, and apathy.[4] Although symptoms generally occur immediately after the simulator experience, their onset can be delayed as much as 18 hours afterward, which may be dangerous for users driving cars or flying aircraft during the time of onset. Symptoms have been found to persist anywhere from less than an hour to as long as 24 to 48 hours.[5] As simulators become more widespread, study of these symptoms will grow in importance. Hopefully these symptoms can be reduced through more realistic simulations, with less lag time and higher resolution, but other problems may arise from long-term use. At present, enough is known about simulator sickness that, in true simulationist form, simulator sickness itself can now be simulated with the help of neural nets.[6]

Psychologically, VR experiences will not be able to simulate certain real world ones. The knowledge that something we are looking at exists before us in physical form influences how we feel about it, a factor which plays a large role in traditional installation art. Multiple objects are used to create an effect in a number of installations, for example, the hundreds of casts of dinosaur bones in Allan McCollum's *Lost Objects* (1991) and the black pictures in frames in his *Personal Surrogates* (1991), or Antony Gormley's *Field* (1991) which consisted of more than 35,000 clay figurines spread across the museum floor. Whereas each physical object has to be created separately, in the computer-generated world of virtual reality graphics can be duplicated thousands of times effortlessly, and so the effect is lost. Site-specific qualities, like light, heat, touch, smell, temperature, air pressure, humidity, and other environmental factors that influence the mood and atmosphere of an installation will not be adequately reproducible in virtual reality. Nor will feelings of physical danger (or the potential for it) be reproducible as they are in physically-based installations. Most museums, of course, would not allow installations which would put their patrons in danger, but certain pieces and installations can at least give the impression of *potential* danger; for example, sharp objects, precariously balanced objects, dark areas which patrons must walk through, not knowing what to expect, performance artists issuing threats of violence, and artworks deliberately designed to be dangerous.[7] The layout of installation art will often include limited entrances and exits, which must be passed through (sometimes in a particular order) and control the way the work is experienced. During a virtual reality session, a user can remove the headset at any point, making every point a possible exit from the work. Installation art, then, will not be threatened by virtual reality, but merely find it complementary, another element to incorporate into itself.

The fact that objects or situations encountered in virtual reality lack the dangers they might otherwise have in the physical world has been used by psychologists in helping patients combat acrophobia (fear of heights),[8] and has pointed at the potential of virtual reality as a tool of study, although there is still no consensus as to how useful it will be. Many of virtual reality's applications, artistic or otherwise, will depend on the nature of the virtual experience needed; depending on how VR is used, its departure from realism can be a strength or a weakness. Just as photorealism is often a goal in computer graphics, reality is the yardstick by which VR's progress can be measured. Though not all forms or applications of VR will necessarily imitate reality, developers of the technology seem to share a design goal of mimicking reality as closely as possible. Much of this enthusiasm is reflected in the terminology used in the VR industry, and the way differences between virtual reality and "real" reality are reflected and effaced.

The term "virtual reality" itself is the best example of industrial optimism; implicit in it is the claim that the difference between VR and lived experience in the material world is negligible; it sets a teleological goal for developers, a continuation of Bazin's "Myth of Total Cinema". The term "virtual reality" began as a brand name at Jaron Lanier's VPL (Visual Programming Language) Research Inc., but quickly came to represent the entire industry in popular usage. Today, the term is applied so widely that there is no industry consensus as to what constitutes "virtual reality"; some claim it must be a networked environment in which multiple users interact, others define it more narrowly as immersion technologies using head-mounted displays, while still others include telepresence and related experiences within the term. The word "virtual", used in optics to describe certain types of images which can not be seen directly,[9] has been used in other ways by the computer industry, such as the "virtual processor" or the idea of "virtual memory", which expands computer memory by using RAM space as if it were hard disk storage space. Apple's *QuickTime VR* software uses the "VR" misleadingly, since it is neither interactive nor a true three-dimensional environment, but rather a 360° image mapped onto a concave surface which corrects perspective as the user rotates through the image. The looseness of the term "virtual reality" has allowed marginally-related products to jump on the "virtual" bandwagon (or the "cyber" one, which is equally vague in popular usage).

While VPL's "virtual" terminology caught on, technologies aimed at recreating multisensory experiences have been around for nearly a century, and have had a long history of development. More than eighty years before simulator rides like Disneyland's "Star Tours", there was "Hale's Tours", which opened in Kansas City on May 28, 1905;

> Another form of specialized motion-picture exhibition that developed simultaneously with the nickelodeons took the viewer-as-passenger convention characterizing many travel programs to its logical conclusion. Hale's Tours and Scenes of the World initiated the craze with a theater that looked like a railway car from the outside. Spectators boarded the "train," paid their dime to a "conductor," and sat in a theater that resembled the interior of a carriage. With rear-screen projection, the film was projected onto a screen at the front of the space—the equivalent of an observation car. In some of the more elaborate shows, the pictures were accompanied by the rocking of the car and the sound of railway clatter. . . . By the following summer, Hale's Tours and its many imitators were popular features at the nation's amusement parks. Claude L. Hagen's "Le Voyage en l'Air," which simulated a balloon voyage, was shown at Coney Island and Happyland on Staten Island, while Tim Hurst's Auto Tours had three cars at Coney Island. Other imitators included the Trolley Car Tours Company and the Trip to California Amusement Company. Such specialized theaters proved to be a fad in the way that the nickelodeons were not. After the 1906 summer season, their impact on film practice receded, although in at least a few situations they continued to operate into the 1910s.[10]

Cinema itself became a more immersive, multisensory experience with the addition of synchronized sound and later the addition of widescreen processes, 3-D, and surround sound. The success of these additions to film prompted some inventors to take the multisensory idea even further. One such pioneer, Morton Heilig, published plans for an "Experience Theater" in 1955, patented a head-mounted stereophonic television display in 1960, and his "Sensorama Simulator" in 1962.

Extending the idea of Cinerama, Heilig's Sensorama was built to simulate the sight, sound, touch, and smell of a particular experience. Howard Rheingold, visiting Heilig in 1990 described the Sensorama experience (minus smell, which was not working at the time);

> I put my hands on the handlebars and rested my face against a viewer that looked like a pair of binoculars with a padded faceplate. Right below the eyepiece was a small grill, near my nose, where the odors would have been pumped in and out of smelling range. Other grills to either side of my face emitted unscented breezes at appropriate times. Small speakers were positioned on either side of my ears. The machine started. I heard an automobile engine, apparently with the muffler removed, saw an expanse of sand dunes, felt my seat lurch, and found myself looking from the driver's seat at a stereographic view of a dune-buggy ride. The film had begun to turn yellowish-brown. It looked as if I were sitting in the front seat and holding on to the handlebars, but I had no way of steering any of the vehicles I found myself riding; I was strictly a passenger. I spent a few loud minutes meandering over sand dunes the size of houses, then I found myself on a motorcycle ride through the streets of Brooklyn, as Brooklyn has not appeared for thirty years.[11]

Other experiments, such as Stan VanDerBeek's "movie-drome" in 1964 and the "Mem-Brain Theater" designed by Rob Fisher, also played with the construction of spaces filled with imagery surrounding the spectator.[12]

Although they are important precursors of virtual reality, Heilig's Sensorama and other early experiments lacked interactivity, now often considered an essential element of a virtual reality experience, and one now possible with digital technology. Computers and interactivity did appear in the work of another pioneer, Myron Krueger, who wrote about "responsive environments" and invented what he called "artificial reality" in the late 1970s and early 80s, years before VPL. Krueger's projects, which included METAPLAY, PSYCHIC SPACE, and VIDEOPLACE, consisted of enclosed rooms with video cameras, wall-sized projected video imagery, and pressure-sensitive modules set into the floor. Interaction occurred as participants stepped on the modules, or as their video image was captured by the camera, manipulated and combined with other computer imagery (sometimes by a human operator), and projected onto the walls, often all in real time.

Both Heilig's Sensorama Simulator and Krueger's Artificial Reality openly acknowledge their unreality; one is a simulator, the other, artificial. Yet neither terminology caught on and developed into an industry, whereas the rather euphemistically-named "virtual reality" did;

as a term, it certainly promised greater mimesis, even if it wasn't as honest as Heilig's or Krueger's terms. Today, thanks to an infusion of VR terminology into popular culture, the word "real" has been given a new shade of meaning; there is "real" as the opposite of "virtual", as opposed to "real" as the opposite of "fake". Despite being opposites to the word "real", the words "virtual" and "fake" are *not* seen as synonymous; while "fake" has a negative connotation, "virtual" has a positive one. "Fake" seems to indicate a copy which lacks positive attributes, for example, a forged painting, while "virtual" seems to indicate a copy lacking negative or harmful attributes (i.e., virtual objects cannot physically hurt a user). The substitution of "virtual" for "fake" as the opposite of "real" is an example of the desire (on the part of either the public, or those corporations selling dreams to the public) to deny the fakeness of certain fantasies, while at the same time it signals a compromise for the less-than-whole experience, one that is "virtually" as good as the real thing —or even better, since the virtual experience is often said to lack certain negative aspects of its real equivalent.

The division between "real" and "virtual" is also complicated by technologies that integrate the two within a single experience. Telepresence, for example, combines a virtual experience on the user's end with data gathered in real time from an actual remote location, which may be physically affected by the user's actions through the robot surrogate present at the location. Another technology combining the two visually is the field known as "augmented reality". Augmented reality, also known as "computer-augmented environments", involves overlaying computer graphics onto a visual image of the physical world. Through the use of multiple video cameras attached to the user's headgear, computers track objects and the user's location. The data is then used to superimpose computer-generated objects or diagrams in such a way that they appear to exist in the same space as the physical objects; the images an augmented reality user receives are a combination of real and virtual worlds (by offsetting the images sent to each eye, dimensional effects are possible). Although they are similar to the heads-up displays used by jet pilots who need to watch their control panels while looking at their immediate environment, augmented reality displays can be more integrated into the image of the physical environment. In one case, augmented reality was used to help a patient suffering from Parkinson's disease;

> Tom Reiss wasn't giving in to Parkinson's disease without a fight. The 46-year-old San Anselmo, California, podiatrist began looking for ways to thwart the increasingly faulty neural transmission lines from his brain to his legs and feet. He'd heard about a Parkinson's-related

> effect called kinesia paradoxa: a trick of mind and eye that enables some Parkinson's patients at certain stages of their disease to momentarily escape the halting, hesitant gait characteristic of their disorder. By following a track of objects placed on the floor at stride-spaced intervals, the effect paradoxically triggers a cycle of movement that imitates near-normal walking.
>
> Riess tried the effect on himself. It worked, but there was a problem: what to do when the track ended.
>
> Riess had heard about specially engineered glasses that let wearers watch TV while, for example, mowing the lawn. If such a device could superimpose the necessary cue of objects on his vision, Riess theorized, perhaps the track could be endless.
>
> That's the question he posed to one of the device's inventors, Thomas A. Furness III, University of Washington professor of industrial engineering. Until now used mostly for entertainment, the device—commercially available from Virtual Visual of Redmond, Washington—projects a big-screen TV image onto one side of a special set of glasses. The image seems suspended in space about ten feet ahead of the wearer and slightly below eye level.
>
> Last May, under supervision of UW research scientist Suzanne Weghorst, Riess tried out the glasses at the Human Interface Technology Laboratory directed by Furness at the Washington Technology Center, located on the UW campus in Seattle. Experimenting with a wide variety of videotaped visual cues—dots, dashes, and other geometric shapes—Riess found he was walking across the laboratory floor at a near-normal gait.[13]

Augmented reality can perform simple functions such as making a video image appear to be some distance in front of a user, but it is quite far from real time interaction with a complex environment. For this to occur, pattern recognition systems will have to reconstruct a three-dimensional version of the visual environment in order to position virtual objects properly—not an easy task. The combination of real and virtual worlds make augmented reality useful for a large number of tasks, since objects can be pointed out or even demonstrated to the user, who can follow superimposed instructional diagrams.[14]

Augmented reality has not yet developed enough to accurately gauge its long-term effects on a user, and it is still far from reaching its potential. Many uses of augmented reality and virtual reality are utilitarian, and while designs and plans are ambitious, the hardware produced is still limited by the available technology, much of which remains prohibitively expensive. And yet that may change, as technology develops. Mirroring, and often inspired by, science fiction writers whose tales of technologically-advanced near-futures have stirred the public's imagination, large corporations and research institutions have built up a body of predictive and speculative work that enthusiastically describes what they see as possible technologies of the near future. While much of this may be hyperbole, technological advances of the past show how even imaginative science fiction writers can underestimate the speed of technological advance. For the most part though, most of the predictions have yet to come true, and existing technology is still lacking in subtlety and riddled with problems. But even if industry is not yet ready to deliver VR into everyday life, it has been successful in delivering virtual reality dreams and fantasies.

Virtual Reality Fantasies

Virtual reality, or technologies similar to it, have appeared in fiction for some time. Dream worlds and fantasy worlds have always existed in literature, but technologically-based ones are more of a twentieth century phenomenon. Notions of virtual reality appear frequently in science fiction stories, from E. M. Forster's *The Machine Stops* from 1909, through writers like Aldous Huxley, Arthur C. Clarke, Philip K. Dick, and finally to writers like William Gibson and others in the "cyberpunk" genre. As computer graphics developed, virtual reality-like technologies became a major theme in science fiction around the late 1970s and into the 1980s, and likewise, serious discussion of what kinds of virtual reality might be possible began appearing in popular discourse.

Rather than focus on the enormous amount of literature concerning virtual reality and philosophically similar situations, we can limit our discussion to those virtual realities which have appeared on film and television, ones which have been visualized and literally "fleshed out" with actors and sets, available to audiences as sound and image (although there is a lot of crossover material; William Gibson's *Neuromancer*, for example, was adapted and performed by the Berkeley Contemporary Opera). Since virtual reality attempts to communicate an experience, film and television are particularly appropriate as they visualize rather than verbally describe their fantasies, making viewer's experiences of them more concrete. Virtual reality, as a technology, varies in the role it plays in these stories: in *Toys* (1993) it barely appears and is used as material for a joke or two; in *Brainstorm* (1983) and *Until the End of the World* (1993) and the television miniseries *Wild Palms* (1993) it is treated more seriously, and as a scientific pursuit; while films like *Tron* (1982), *Total Recall* (1990), *The Lawnmower Man* (1992), *Virtuosity* (1995), *The Matrix* (1999), and the short-lived television series *V. R. 5* (1995), fantasize more and give virtual reality a central place, with important action and struggles for control occurring within virtual spaces. In many of these dramas, much of the narrative conflict is generated within the virtual world, giving it a dystopic pallor, or around the controversial nature of the technology itself in the real world; VR is a contested technology producing a contested virtual space.

Characters who enter virtual reality worlds in these tales often have more power there than they do in the real world (Case, the main character of *Neuromancer* who becomes depressed when he can no longer "jack in", is a prime example). In virtual reality, these characters have power in the fantasy world itself, or through it, telepresent power in the real world. Again, the notion of a "space inside the computer" is essential to its visualization in these stories. In *Tron*, the virtual world is not so much one generated by the computer as is itself the "inside" of the computer itself, where programs and subroutines are represented by people and vehicles. The world is ruled by the evil Master Control Program, who also runs many things in the outside world, and can only be stopped (it seems) by "going into" the computer. Likewise in *The Lawnmower Man*, power and control emanate outwards from the computer and the network of technology that it controls, prompting the title character to "go into" the virtual world and assume command. In *Ghost in the Machine* (1993) a killer is somehow "zapped into" a computer system, and for the rest of the film travels along electrical lines and controls various forms of electric and electronic technology. Even in *Until the End of the World*, and especially in *Brainstorm* and *The Matrix*, virtual reality becomes a powerful tool which characters compete to control, and in *Blade Runner* and *Total Recall*, implanted memories and experiences are used as a means of control. Virtual reality, then, is depicted as the place where power resides; as if it were more another dimension instead of mere fabricated imagery.

In many of these films, characters not only use virtual reality, but actually exchange their physical bodies for digital ones, effectively "entering" the computer instead of merely using it, making virtual experiences "real" for them. In *Ghost in the Machine*, (with by far the weakest narrative explanation) a killer gets "zapped" into electronic form during an electrical storm, and ends up terrorizing his victims from "inside" the technology, while in *The Lawnmower Man*, the title character gets sucked into virtual reality, the datasuit he is wearing deflating like a balloon as he is pulled in. *Tron*'s far-fetched solution to this problem is to have its main character digitized in by a laser which scans three-dimensional objects (disintegrating them in the process) and then rebuilds them "inside" the computer in cyberspace.

At the end of the film, the process is reversed, and laser rebuilds the main character back in the physical world.[15] Ironically, by physically entering the computer, characters become more vulnerable; they may be destroyed there never "get out" again. But if they were merely telepresent, or still existent outside the computer, there would be less danger to them, and narrative suspense would be lost. Likewise, the virtual world they inhabit becomes more real, and is without the always-available exit points found in virtual reality wherein the user can remove the head-mounted display at any point or leave the simulator. Thus, narrative needs have also resulted in an overvaluing of the technology and an exaggeration of its possibilities.

The idea of scanning, which "records" form by quantizing an image or object into 2-D or 3-D grids of elements that are represented electronically and stored sequentially, is the key metaphor used in connecting physical and virtual worlds. In "matter transmission" fantasies, people become objects, mere grids of matter that can be turned into information and transmitted. Perhaps the most well-known use of this imaginary technology are the transporters used in all of the *Star Trek* television series. People are "beamed" up and down from planets, across space, ship to ship, and can even be stored in the computer's memory if needed. Conceptual problems —for example, whether the person stored in memory could be rematerialized in multiple copies— are rarely, if ever, addressed.

The *Star Trek* universe is the most detailed in science fiction, logging in hundreds of hours of television episodes from several series and an increasing number of feature films. It contains a level of detail that extends beyond the shows into books, computer software, and fan conventions.[16] Building on the technology begun in the original series, *Star Trek: The Next Generation* combined transporter technology, force-field technology, and advanced computing technology in *Star Trek*'s most ambitious creation yet: The Holodeck. The Holodeck is a form of virtual reality which takes place a large empty room, where yellow grids line the walls when the holodeck is turned off. Through a combination of holograms and force fields, the computer can create, at a character's command, a complete environment with what appear to be solid objects. Holograms make the objects visible, while force fields give them tangible form. Characters can go mountain climbing or whitewater rafting, and the computer can even generate characters who are a part of the fantasy world. Since it is completely interactive and wholly immersive with a transparent interface, the Holodeck could be said to be the logical goal of the virtual reality industry. Indeed, even the scholarly reference work *Virtual Reality: An International*

Directory of Research Projects opens with an assessment of "the Ultimate Display" using the Holodeck as an example;

> As a display device, the Holodeck is miraculous. Its users have a visual display that produces a 360° field-of-view image. . . The image is truly stereographic for all users from any viewpoint. All distance cues are present, including focal accommodation. Scene complexity is very high, and is indistinguishable from real world scene complexity. The users do not need to wear HMDs to see the visual images.
>
> Auditory information is spatially localized to the source, with environmental reverberation to improve accuracy. Sound generation, which includes speech and non-speech, is personalized to each individual, even though multiple species with different head and ear structures are present. Again, no equipment needs to be worn in the Holodeck.
>
> The haptic display capabilities are equally amazing. Users have full proprioceptive and somatic feedback consistent with the environment. Interactions with objects in the environment produce appropriate limitations of proprioceptively sensed motion. If you push against a wall, you feel your arm stop as well as see your arm stop. This is accomplished without the use of exoskeletal devices. Somatically sensed stimuli are also consistent with the environment.[17]

Instead of turning physical characters into digital entities as in *Tron*, *The Lawnmower Man*, and *Ghost in the Machine*, the virtual environment of the Holodeck makes the virtual appear physical, completely immersing its users in what is experientially similar to a physical environment. The illusion is so complete that in one episode, "The Inner Light", Captain Picard wakes up in an unusual place where people he has never seen before pretend to know him, and immediate thinks he is in the holodeck. And scenes set in the holodeck are not always identified as such immediately; often the scene appears real until something gives the scene away as a simulation, for example when the action freezes or background characters vanish.

The Holodeck's illusions are so tangibly real due to the force fields that users can even be injured by virtual objects; for example, in *Next Generation*'s Episode 12, "The Big Goodbye", the first episode in which the holodeck appears, one of the characters is injured by a bullet. In other episodes like "Fistful of Datas", they can face mortal danger when programs malfunction or the computer fails to respond to the spoken commands that usually give users complete control over their virtual world. This sense of possible danger is a strategy, like that having the user "enter" the computer, which raises the stakes, since the user may not be able to leave the holodeck the way one can leave a simulator or remove a head-mounted display. On numerous occasions, including the first two episodes to use it,[18] the Holodeck is malfunctioning, either through some character's deliberate actions, some cosmic energy wave or disturbance, or computer malfunctions. On such occasions, characters must fend for themselves against virtual adversaries until such time as the malfunction can be repaired. Besides making any number of exotic locations available, the Holodeck also brought with it some interesting —though far-fetched— possibilities; for example, in one episode, "Elementary, Dear Data", a user error on the Holodeck "results in accidental creation of a computer software-based sentient intelligence within a simulation program".[19]

The series *Star Trek: Voyager* takes Holodeck technology a step further, with their emergency medical holographic doctor, who is intended to be a backup but ends up serving as a full-time doctor when *Voyager* becomes lost in space. The doctor combines Holodeck technology with a gigantic medical database, and with his "mobile emitter", he can even walk freely about the spaceship. He is, for all intents and purposes, a member of the crew, an idea pointed out in the episode "Eye of the Needle". The doctor even uses the holographic/force field technology in his practice; in "The Phage", Neelix's lungs are stolen by an alien, and in order to keep him alive, the doctor places him in a restraint, and creates a holographic lung functioning within his body until a real one can be found. Once again, the virtual has replaced the real; every advance of *Star Trek* technology seems to reduce the differences between them. By using its technology in sickbay, Holodeck technology is no longer limited to a separate, recreational area of the ship, it becomes a life-sustaining technology.

The Holodeck, when not in use, is a large empty room, with a yellow grid on the floor, walls, and ceiling; and when in use, it can become any location the characters want. The Holodeck can be seen as a metaphor for a television or movie studio space, wherein sets are built and disassembled, and spaces are created and inhabited by characters. Often in the Holodeck, mini-narratives occur, in which crew members are dressed in the appropriate costumes; as gangsters, detectives, swashbucklers, cowboys, and so on. Since all the locations on the *Enterprise* or *Voyager* are sets built in a studio, the Holodeck is a set within a set; it is literally a set as well as a metaphor for one. All events occurring within it are no less real than those on the other sets; the differences between them lie entirely within the diegetic world. An irony here is that while in most cases the set —the fake— is made to be real (at least to the characters), objects on the Holodeck, which are physically (nondiegetically) real, are taken to be virtual within the diegetic world!

The Holodeck is also a metaphor for the television set itself; a small cubical space whose "interior" —what we see on-screen— can play out any fantasy, representing anything. Even more interesting, the television (often found in the family room, with (typically) the couch and chairs set facing it), and the family room itself, are also mirrored within *Star Trek: The Next Generation*. The central location of the series is the bridge, which bears a striking resemblance to a family room with a TV in it; in the front of the bridge there is a large, wall-sized viewscreen, on which adventures often unfold, and through which contact is made with the outside world, while throughout the room there are consoles and chairs, where crew members (the *Star Trek* "family") sit facing the viewscreen. (We could even see television as producing "hailing frequencies" enticing viewers to "open a channel" and watch.) But, here, too, VR fantasies enter the picture; while events may be as mediated for the TV viewer as they are for the *Star Trek* crew, the TV does not allow anything near the kind of interaction that the crew has with *their* viewscreen.

The bridge, then, is in a kind of virtual space; the viewscreen and incoming frequencies are the main link to the universe outside, which is entirely mediated for the crew by the ship itself; technology *is* their environment. The bridge, as an enclosed space with screen and seats, is also like a simulator ride, and, as any *Star Trek* fan knows, often shakes like one. As the TV viewing audience watches the characters on the bridge, experiencing the *Star Trek* world vicariously through them, watching the bridge shake becomes a *virtual simulator ride*, a simulator ride which we occupy in only a virtual sense. As the Holodeck is a set within a set, watching the bridge shake is a simulation of a simulation, as the "shaking" of the bridge is itself simulated by a camera shake and stumbling actors.

Despite the virtual nature of the *Star Trek* universe, some of it crosses over into the real world; the *Enterprise* is indeed a real spaceship. As Jerry Grey writes, "It is no coincidence, incidentally, that *Enterprise* is the name of Captain Kirk and Mr. Spock's *Star Trek* spaceship and of the first *real* U. S. spaceship, Shuttle Orbiter No. 101..."[20] On an episode of the PBS series *The New Explorers* entitled "The Science of Star Trek", other parallels between *Star Trek* and the real world are mentioned. For example, certain NASA employees proposed that NASA adopt the *Star Trek* motto as its own; SACI, a defense contractor, wanted their command console to look like the Enterprise's on *Star Trek*; and the Dante project's telepresence robot plots its spatial information on a black background with a yellow grid on it, the design taken from the Holodeck interior. The program also reveals that Mae C. Jemison, the first black woman astronaut who rode the space shuttle *Endeavor*, was inspired by *Star Trek* as a child, and later appeared in an episode of *Star Trek: The Next Generation* in a cameo role.[21]

As such examples show, *Star Trek* has been an influence not only on the imagination of the public, but on NASA's space program as well. Science fiction has often acted as a source of names and concepts later developed into real technologies, as imagination paves the way for invention. Currently in the world of virtual reality technology, there is far more imagination than invention; companies developing VR are selling a dream as much as they are hoping to sell a product. But so much of what we hear about virtual reality amounts to plans still on the drawing board, surrounding by promise and hype, existing only in the imagination; what Ellen Strain has called "'virtual' virtual reality".[22] Even scholarly literature is not immune from fantasizing; after giving ten possible definitions of cyberspace, Michael Benedikt begins the introduction to his anthology *Cyberspace: First Steps* by stating "Cyberspace as just described—and, for the most part, as described in this book—does not exist."[23] Apart from military simulators and work going on in research laboratories, the virtual reality available to the public is mostly recreational; virtual reality video games like Virtuality's *Dactyl Nightmare* (1992), multi-player video games like Virtual World's *Battletech* and *Red Planet* or Visions of Reality's *Cybergate*, and around thirty different simulator rides, located at theme parks and tourist attractions across the country. These rides are still very expensive (often around a dollar a minute) and are mainly in urban centers where better profits are ensured. The other strand of commercially available virtual reality is computer software, like *Autocad*, *Virtus Walkthrough*, or Straylight's *PhotoVR*, for building virtual worlds; for a moderate investment, VR consumers can become VR producers, and VRML (virtual reality markup language) may soon join HTML in the creation of internet websites. Increases in computing power and software sophistication is slowly allowing home-brewed VR to grow into a cottage industry, in the same way video games and computer graphics have.

For the moment, the virtual reality that exists will have to act as a temporary substitute for the virtual reality dreamed about in fiction. As mentioned earlier, "virtual reality" need not be limited to the immersion rides or devices of industry, but rather a philosophical condition resulting from mediation, abstraction, and the trading of direct linkages to the material world for more indirect ones.

Acceptable Substitutes? Displacements and Replacements

New technologies often do not *replace* old technologies so much as they *displace* them. When photography appeared, painting was not dead as many feared; it merely developed in a different direction. Film did not replace theater, nor did television replace film; each medium is still around, in its own niche. Even the telegraph is still in use today, and the post office is busier than ever. But how much *dis*placement is needed to become *re*placement? A technology may not completely vanish, but its niche may grow so small that it becomes little more than a curiosity or museum piece, like the optical toys of the nineteenth century, mechanical televisions, vacuum tube appliances, or even early home computers and games like *Pong*. And it is not just a matter of new technology replacing old technology; as many authors have been pointing out over the last decade or so; technology has often displaced or replaced experiences rooted in the natural world.[24]

The "virtualizing" of reality, then, can occur overtly or covertly; while people openly consider simulator rides, telepresence, and head-mounted displays to be "virtual", they generally do not consider an answering-machine message or the evening TV news to be "virtual" at all, even though they may be no less technologically based. Even the notions of 'overt VR' and 'covert VR' do not cover all the ambiguities, for most forms of VR lie somewhere in between. Perhaps the most overt virtual reality experience is that of head-mounted displays, since they are very noticeable (and sometimes painful) and because their worlds are (still) far from resembling unmediated reality. Games qualify as well here, since their situations are often obviously fictional. Ride-in simulators are overt as well, but slightly less so; the body's senses can be fooled, although the mind may not be. Simulators used for training (e.g., for pilots) must have some degree of realism, simulating situations could actually occur; that is the idea behind a training simulator in the first place. Simulators and computer simulations, as well as movies, television, and telepresence, are a mixture of overt and covert "virtual" reality experiences; each mixes fantastic or hypothetical elements with ones taken from the real, material world of lived experience, to different degrees. In almost every single example of these media, elements of both fantasy and reality can be found.[25] Still less acknowledged as virtual experiences are such things as home video, photographs, or answering machine messages (or for that matter, "live" telephone conversation itself). In these cases, the content does not originate with a corporation supplying entertainment, nor is such content considered entertainment; but the use of such media still constitutes a virtual, and indirect, experience. With more and more mediation, our sense of the world becomes more of a virtual experience; the real world becomes more virtual. Instead of wondering how to keep virtual and real worlds separate, perhaps we should be asking how we can keep our experience of the real world from become virtualized. Forms of overt VR, such as games, simulator rides, and so on, are far from becoming convincing replicas of physical world experience (and the Holodeck will never materialize), and so they need not become a worry as yet. Mediated substitutes for reality, however, are far more common, and consist of technologies that have become abstracted, invisible, and part of the background of our everyday experience; they are no longer in the spotlight the way VR and simulators are, and hence go relatively unnoticed. In this sense, overt VR makes covert VR seem more real by comparison. The technology making up the digital environment, mediating late twentieth century reality, is the ultimate in technological immersive experiences.

Substitutes are accepted everywhere; mediated experiences are accepted in place of real ones, despite their much narrower experiential bandwidth, their lack of context, and their machine-controlled or third-party mediation. Within media, there are substitutes of substitutes; compressed "lossy" images used in place of uncompressed lossless ones, and lower resolution versions of high resolution images or sounds. A user who experiences something for the first time through a substitute —as often happens— has no standard by which to compare the experience; without knowledge of the actual experience or thing itself, the user of a substitute will not even realize what is being missed, or how far the substitute falls short of what it is supposed to replace or represent. Thus the substitute not only takes the place of the original, it can *become* the original, an idea central to many postmodernist works of art exploring these themes through reappropriation and imitation.

Why are substitutes so readily and widely accepted? What do substitutes provide that the original cannot? The answer often has something to do with some form of *cost*, which may be measured in access, control, availability, convenience, money, computer memory, time, comfort, effort, or risk. The substitute, of course, generally has less value than what it replaces (in some sense, otherwise it would not be merely a substitute); but it makes up for this difference by costing less (in some *other* way) than the original would have (reflecting the reduction in value in what appears to be a positive way). But substitutes vary in their approximation of the quality of the original; a substitute of lower quality will also have a certain *cost* to the user, since a substitute, by definition, has less value than does the original.

Thus the user of a substitute loses something (experientially speaking) in the process of using a substitute at the same time as some other cost (money, time, effort, etc.) is reduced. The substitute may be of such poor quality, that the cost of the experiential loss outweighs the initial cost (measured in money, time, effort, etc.); at this point the substitute becomes unacceptable. But for the condition of unacceptability to occur, *one has to have a sense of the experiential loss in the first place*; and this sense of the original, real and direct experience, is something that the user of a substitute may be lacking. How can one judge the quality of a substitute without knowledge of what it replaces? Without any criteria to judge by, a user of a substitute may be less likely to deem it unacceptable. The substitute that a person is accustomed to, or has always used, then becomes the original, in that person's experience, setting the standard rather than having to measure up to one. As the chain of substitutes replacing one another continues to worsen they will still have to measure up to this standard, but they, too, may end up replacing it, as the cycle repeats.

The appearance of a substitute implies that the real thing is too costly in some way; too expensive financially, too time-consuming, too difficult or dangerous, (or in the case of foods, too fattening) and so forth. While there may be good reasons to apply standards in some matters, such standards are often set arbitrarily, either by commercial advertisements or cultural values, and can be especially treacherous when they apply to the quality of life; who is to say that reality is inconvenient, or threatening, or takes too much effort? Substitutes, then, can change perception by pointing out costs and alternatives; certain foods widely enjoyed in the nineteenth century are now looked upon with horror by some people, and today most people consider the cost of not having an automobile to be greater than the cost of having one. A substitute's acceptability or unacceptability indicates an underlying value system, determined by what certain tradeoffs are worth. The Amish, for example, do not agree with many of the technological tradeoffs that most of Western society has agreed upon. Traditional and tribal cultures likewise have different attitudes about such tradeoffs, and cultures differ from one to the next in what is considered an acceptable substitute (for experiences and conditions as well as material goods).

As people, and entire cultures, give up certain direct or "real" experiences, displacing or even replacing them with what they consider to be acceptable substitutes, their indexical connection to the outside world may grow more tenuous and distorted. Much of what they know will rely more on the reports of others (mostly strangers) rather than their own firsthand experience. Strings of replacements and substitutes will dilute experience until a belief in the possibility of an original or authentic experience are gone (some would argue that this has already happened). As these series trade-offs continue, there may soon be no standard with which to measure an experience, no common ground to make comparisons by, and even no index by which one can measure the loss of indexicality.

NOTES

1. Benedikt, Michael, "Cyberspace: Some Proposals", in Benedikt, Michael, editor, *Cyberspace: First Steps*, Cambridge, Massachusetts, and London, England: The MIT Press, ©1991, page 124.
2. Heim, Michael, *The Metaphysics of Virtual Reality*, Oxford and New York: Oxford University Press, ©1993, page 117.
3. Gene Youngblood, as quoted in *The Computer Revolution and the Arts*, edited by Richard Loveless, Tampa, Florida: University of South Florida Press, ©1989, page 13.
4. On simulator sickness, see Greenfield, Richard, "Simulator sickness: virtual reality", *Computer Weekly*, March 17, 1994, page 38: Regan E. C., and Price K. R., "The frequency of occurrence and severity of side-effects of immersion virtual reality", *Aviation, Space, & Environmental Medicine*, June 1994, Volume 65, Number 6, pages 527-530: and Crowley, John S., "Simulator sickness: A problem for Army aviation", *Aviation, Space, & Environmental Medicine*, April 1987, Volume 58, Number 4, pages 355-357. Also see Voge, Victoria M., "Simulator sickness provoked by a human centrifuge", *Military Medicine*, October 1991, Volume 156, Number 10, pages 575-577: Hettinger, Lawrence J.; Berbaum, Kevin S.; Kennedy, Robert S.; Dunlap, William P.; *et al.*, "Vection and simulator sickness", *Military Psychology*, 1990, Volume 2, Number 3, pages 171-181: and Uliano, K. C.; Lambert, E. Y.; Kennedy, R. S.; Sheppard, D. J., "The effects of asynchronous visual delays on simulator flight performance and the development of simulator sickness symptomatology", *US Naval Training Systems Center Technical Reports*, December 1986, Number 85-D-0026-1.
5. Greenfield, Richard, "Simulator sickness: virtual reality", *Computer Weekly*, March 17, 1994, page 38; and Voge, Victoria M., "Simulator sickness provoked by a human centrifuge", *Military Medicine*, October 1991, Volume 156, Number 10, pages 575-577. Also see Baltzley, Dennis R.; Kennedy-Robert-S.; Berbaum-Kevin-S.; Lilienthal-Michael-G.; *et al.*, "The time course of postflight simulator sickness symptoms", *Aviation, Space, & Environmental Medicine*, November 1989, Volume 60, Number 11, pages 1043-1048.
6. Glenn O. Allgood has devised a neural network paradigm that can predict of onset & level of simulator sickness. See Allgood, Glenn O., "Development of a neural net paradigm that predicts simulator sickness",

Dissertation Abstracts International, December 1991, Volume 52 (6-B) 3336.

7. An example of artwork deliberately designed to be dangerous can be found in Raty, Panu, "The Helsinki Killer Ball" in the "Electric Word" column in *Wired*, July, 1995, page 39. Describing the emormous steel ball, Raty writes, "It has a motor, motion sensors, a juicy battery, and the mind of a psycho killer. It hunts you. If you don't jump aside, you'll be trampled under it." Obviously there is a danger here which virtual reality would have a hard time duplicating.

An example of a performance artists issuing threats is an installation piece by Vito Acconci. As Kathy O'Dell describes it;

> Perhaps no other piece from the early 1970s more thoroughly spells out the psychologized drama engendered by performance-based video than Acconci's *Claim* (1971). Blindfolded, seated in a basement at the end of a long flight of stairs, armed with metal pipes and a crowbar, threatening to swing at anyone who tried to come near, Acconci simultaneously invited and prohibited every visitor to the 93 Grand Street loft to descend into the world of the unconscious.

See O'Dell, Kathy, "Performance, Video, and Trouble in the Home", in *Illuminating Video: An Essential Guide to Video Art*, edited by Doug Hall ad Sally Jo Fifer, New York: Aperture Books in association with the Bay Area Video Coalition, ©1990, page 136.

8. See Mahoney, Diana Phillips, "High expectations for virtual therapy", *Computer Graphics World*, Volume 17, Number 9, September 1994, page 14(2); and Goddard, Alison, "Virtual therapy reaches new heights", *New Scientist*, 142, June 11, 1994, page 6. Virtual reality in general has yet to live up to the hype surrounding it; see Gibbs, W. Wayt, "Virtual Reality Check: Imaginary environments are still far from real", *Scientific American*, December 1994, page 40.

9. A "virtual image" is described in *The Focal Encyclopedia of Photography* as being,

> The image seen after refraction by a diverging lens or after reflection in a plane or convex mirror when the light rays diverging from an object point cannot then be brought to a corresponding image point. The virtual image is seen by looking at the subject via the mirror or through the lens at the subject and cannot be received directly on a screen or film surface.

From Ray, Sidney F., "Virtual Image", in Stroebel, Leslie, and Richard Zakia, *The Focal Encyclopedia of Photography*, Boston, Massachusetts, and London, England: Focal Press, ©1993, page 840. A virtual image, then, cannot be seen directly, unlike a real image which can be. "Virtual" then, seems to imply an additional layer of instrumental mediation.

10. Musser, Charles, *The Emergence of Cinema: The American Screen to 1907*, Charles Scribner's Sons, New York, ©1990, pages 429-430.
11. Rheingold, Howard, *Virtual Reality*, Summit Books, Simon & Schuster, Inc., New York, ©1991, pages 52-53.
12. See VanDerBeek, Stan, "Re-Vision of Cine-Dreams", *Dreamworks*, Fall 1981, pages 4-12.
13. From the PR Newswire report ""Virtual Track" for Parkinson's Disease Sufferers", *IEEE Computer Graphics and Applications*, March 1994, page 90.
14. An example of this, in which computer overlays guide a user in the repair of a copier, can be found in MacIntyre, Blair, and Seligmann, Doree, "Knowledge-based augmented reality", *Communications of the ACM*, July 1993, Volume 36, Number 7, page 52(11).
15. As if anything could be rematerialized or reconstructed by a laser. Matter transmission, in which matter is broken down into energy, made into a signal, transmitted, and then rebuilt back into physical objects is a standard technology is science fiction. Early appearances can be found in the stories "The Man Without A Body"(1877) by Edward Page Mitchell and "Professor Vehr's Electrical Experiment"(1885) by Robert Duncan Milne, and later examples include *The Fly*(1957, remade 1986), Stephen King's short story "The Jaunt", and the current transporter technology of the four *Star Trek* television series, which allow characters to "beam up". From the entry on "Matter Transmisson" in *The Encyclopedia of Science Fiction*, edited by John Clute and Peter Nicholls, New York: St. Martin's Press, ©1993, pages 787-788.
16. On June 13, 1994, a picture of a Klingon even appeared on the front page of *The Wall Street Journal*, with an accompanying article on how there were several teams of researchers debating how the Bible should be translated into the Klingon language. According to the article, there is even a nonprofit Klingon Language Institute which is translating all of Shakespeare's works into Klingon. See Dolan, Carrie, "Translating the Bible into Klingon Stirs Cosmic Debate: Some Favor a Literal Tack, Others Find That Alien: Help From the Lutherans", *The Wall Street Journal*, June 13, 1994, First page and page A4.
17. Discussion of the fictional Holodeck is separated of from the scholarly text by the following paragraph entitled "Caveats";

> This example is not intended to address in detail research in virtual environment technology. Just as Richard Hamming said that "the purpose of computing is insight, not numbers", so the authors wish to give the reader an insight into the problems facing the VE research community. Concepts introduced in this section are explained more fully later. Also, the detailed characteristics of the Ultimate Display described in this section are products of the authors' imagination, with due thanks to the creators of *Star Trek*.

From Thompson, Jeremy, *Virtual Reality: An International Directory of Research Projects*, London: Meckler, ©1993, page 4.

18. Episode 12, "The Big Goodbye"; and Episode 15, "11001001", in which the Binars keep Picard and Riker distracted inside the Holodeck while they hijack the Enterprise.

19. Okuda, Michael, and Denise Okuda, *Star Trek Chronology: The History of the Future*, New York: Pocket Books, ©1993, page 106.

20. Grey, Jerry, *Enterprise*, New York: William Morrow and Company, Inc., ©1979, page 12. With the *Enterprise* named after the *Star Trek* ship, one wonders if the space shuttle *Discovery* was named after the spaceship in *2001: A Space Odyssey*.

21. From "The Science of Star Trek", an episode in the PBS series *The New Explorers*, produced by Kurtis Productions Ltd. for WTTW, and broadcast Wednesday, January 18, 1995, on KCET, Los Angeles, at 9:00 p.m. Pacific Standard Time.

22. From Strain, Ellen, "Mastering New Worlds: Tourists in Virtual Reality", a paper given at the *Visible Evidence II Conference*, 1994, at the University of Southern California, August 18-21, page 1.

23. Benedikt, Michael, "Cyberspace: Some Proposals", in Benedikt, Michael, editor, *Cyberspace: First Steps*, Cambridge, Massachusetts, and London, England: The MIT Press, ©1991, page 3.

24. Authors such as O. B. Hardison, Neil Postman, Lewis Mumford, Jacques Ellul, and even Henry David Thoreau have been writing of the abstraction from nature through technology, and films like Godfrey Reggio's *Koyaanisqatsi* (1983) and *Powaqqatsi* (1988) also provide interesting commentaries.

25. No mediated reality is ever objective, nor is every element of it completely known, thus elements of speculation, or fantasy, creep in. Likewise, the most fantastic story will have some basis or tie in to the physical world we know, even if is merely a sense of cause and effect based on laws of physics.

10.
Indexicality

As communication grows more mediated, technology pervades and reshapes the environment, and as the virtual and real are combined in a variety of ways, our relationship to the world around us changes; an abstracting of reality occurs. In order to more closely examine how digital technology affects experience, it is useful to turn to semiotics, the study of signs and meaning. Meaning is conveyed by signs, and signs can be related to what they represent in different ways. Semiotician Charles Sanders Peirce divided all signs into three categories;

> A sign is either an *icon*, an *index*, or a *symbol*. An *icon* is a sign which would possess the character which renders it significant, even though its object had no existence; such as a lead-pencil streak as representing a geometrical line. An *index* is a sign which would, at once, lose the character which makes it a sign if its object were removed, but would not lose that character if there were no interpretant. Such, for instance, is a piece of mould with a bullet-hole in it as a sign of a shot; for without the shot there would have been no hole; but there is a hole there, whether anybody has the sense to attribute it to a shot or not. A *symbol* is a sign which would lose the character which renders it a sign if there were no interpretant. Such is any utterance of speech which signifies what it does only by virtue of its being understood to have that signification.[1]

Peirce defined the index by setting it in relation to his other two types of signs, the icon and the symbol. On the one hand, the icon resembles what it represents and would remain significant "even though its object had no existence"; the icon thus has some independence from its referent. The icon may represent an *object*, such as the pedestrian represented on a "Ped Xing" sign, or it may represent a *process*, such as the trash can icon found on the Macintosh computer screen. The trash can icon does not represent some trash can in the physical world, but rather the process involving the trash can, that of throwing something in the trash. The symbol, on the other hand, is arbitrarily assigned, and is thus completely dependent on the interpreter of the sign to make sense of it. The index falls somewhere in between the two; it is neither arbitrary, nor is it completely independent of its referent. It maintains a physical link or kinship to its referent, and its own existence depends on the existence of its referent. While the icon represents a generalization (in the sense that it has meaning without a specific referent), and the symbol tells us little without the symbolic system of which it is a part and which it is defined by, the index is specific in its representation, and in a manner which need not rely on a symbolic system (or interpretant). One of Peirce's examples of an index was a weathervane which signified the direction of the wind because of its physical relation with the wind. More than either the icon or the symbol, the index is a direct link to the thing in the real world that it represents. In the previous chapter we defined "real" as the *least mediated* (or *im*mediate) experience available to an individual, the content of direct experience; and so the construction of "reality" in the mind of the individual is dependent on indexical connections. Indexicality, then, concerns the nature of these connections, which are interpreted by the individual as sensory information.

Senses and Signals

Sight, sound, smell, taste, and touch are ways in which human beings can experience their immediate environment. For every sensory stimulus, there is a physical transaction with the environment; the body is in contact either with photons, sensed as light (or heat), sound waves traveling through some medium, or the physical matter involved in taste, touch, and smell (which detects trace amounts of a substance). The body is also in contact with things it cannot sense directly; cosmic rays, x-rays, gamma rays, and other high frequency radio waves, for example, pass through the body undetected. One band of radiation frequencies, the visible spectrum of light, can be detected by the eye; and those immediately bordering the visible spectrum are detectable only in large amounts (infrared radiation may be felt as heat, and ultraviolet radiation will produce sunburn). Sound likewise has a limited range of frequencies which are audible. And all senses are limited by distance, especially taste, touch, and smell.

Sensory data, as indicators, are composed of indexical signs, and can be thought of as signals carrying information. Media are used to extend this sensory range, augmenting it with technology that translates an index into imperceptible electronic signals, and then translates those signals into perceptible ones. In electronics technology, devices known as *transducers* convert signals from one medium to another. Microphones convert acoustic signals into electronic ones, while loudspeakers convert electronic ones into acoustic ones. Video cameras and television monitors perform similar functions for both sound and image. Some transducers, like motors or generators, can convert mechanical movement to electronic signals and vice versa. Phonograph cartridges, CD players, seismographs, strain gauges, and Geiger counters are other examples of transducers. Transducers, by converting imperceptible signals into perceptible ones, allow a variety of different indexical connections to be made to otherwise imperceptible referents, extending the human sensory range by artificial means. Often, because there are transducers at both ends of a connection, the transducers' function appears transparent; for example, when talking to someone on a telephone, most people think of the voice they hear as the person at the other end, instead of an acoustic signal reconstructed from an electronic one.

Although sensory transducers are designed with transparency in mind, they are never completely transparent, and the nature of this perceptual discrepancy differs from one media to another. For example, besides the differences in size, resolution, and viewing conditions, the film screen and video screen differ even at the level of the eye; projected film creates color subtractively, while video creates it additively. The film projects colors which already exist on the film, whereas every pixel on the television screen is made up of three small circles of light (red, green and blue) set in very close proximity and glowing at various intensities; because of the proximity, the eye blends these colors together to produce the desired colors, in a manner similar to techniques used in impressionist painting. Thus, the color production is done by the eye as opposed to film where the color is mixed before reaching the eye. There are fewer pixels available to create an image, which means there is a greater reliance on Gestalt psychology principles to "connect the dots" into a meaningful image. However, because of the advantages and flexibility that video and computer monitors have over projected film imagery, such differences are usually seen as being trivial.

As Walter Benjamin demonstrated in his essay "The Work of Art in the Age of Mechanical Reproduction", mechanical reproduction changed not only works of art, but the definition of Art itself, and its place in society. In a similar manner, electronic imaging and digital technology change the indexical status of the image and its uses, as well as notions of indexicality. The media's epistemological claims, and particularly those of non-fiction reportage, are dependent on the indexical nature of images and sounds, and the value that indexicality has for the spectator. Epistemological claims with an indexical basis can be grouped into two categories, one involving the information attained, and the other involving the process of attainment. First, the camera and the lens are designed so as to reproduce the rays of light converging on a point in space from a particular angle of vision and direction. Even when the focal length does not match that of the eye and color is reproduced as a grayscale, the image is still valued as evidence for its ability to visually reproduce a scene. The second claim made, and emphasized by Bazin, is the "objective" nature of the image due to its having been produced by a chemical reaction on the film, without human intervention. Bazin seems to value this claim more than the first when he says, "No matter how fuzzy, distorted, or discolored, no matter how lacking in documentary value the image may be, it shares, by virtue of the very process of its becoming, the being of the model of which it is the reproduction; it *is* the model."[2] Both of these claims are carried over into digital technology, which uses a lens system similar to a camera, and which produces an image algorithmically (although electronically instead of chemically). But here the similarities end.

A Grain of Truth?
Changing Values of the Digital Image

As a means of documentation, the photographic image is valued for its attempt to produce a record of the light rays converging on a single point in space, if nothing else; but digital composites can be clean and seamless and images can exist without referents. Compositing, of course, has always been around, as well as other forms of photographic deception, but digital technology makes it easier —and more widespread— than it has ever been. Unlike media in which an image is inscribed in tangible form upon a surface, the digital image exists as an ethereal display derived from a set of numbers. Formal aspects of the image which normally indicate tampering, such as inconsistencies in color, lighting, focus or resolution, and matte lines or overlaps, can be controlled and effaced, leaving no incriminating evidence of the alteration. The ease of manipulation, sometimes used for purely aesthetic purposes, is a temptation hard to resist; even *National Geographic*, with its emphasis on documentary value, has used photographic restructuring. On the cover of the February, 1982 issue, the background pyramid was moved in closer to a foreground one to produce a better composition, with no notice of this distortion given. Most such deceptions, like this one, have come to light through information beyond the image itself; reports, admissions, and so on. However, without such additional information to indicate such changes, many deceptions would never be uncovered.

Even when attempts are made to capture images as unaltered and accurately as possible, transformations occurring at the image-capture stage affect an image's status and indexical value. Unlike film photography where a latent image is formed by a chemical reaction, digital cameras actively employ technology in recording an image, translating it directly into an electronic signal. In the process of quantization, each image is broken up into a grid of discrete single-color picture elements, or *pixels*, which produce the image when viewed together. An image's colors are likewise limited to the gradations of color available in the process.

Quantization, then, unavoidably affects the image by limiting its spatial, temporal, and color resolution, in order to reduce the image into readable machine code of ones and zeroes that indicate the levels of pixel values. Thus images become limited to the resolutions (and frame rate) recorded at their initial capture, as well as dependent on the algorithms and hardware needed to reconstruct them, and the digital grid which each picture is mapped onto will have its inevitable effects on the images (some of which have already been discussed previously).

Spatial resolution is the measure of the number of pixels that make up the image. How these pixels are manipulated will determine the texture of the image; the picture could be translated from the grid into hard-copy or other media as grain, dither, half-tone, or a lower resolution grid. Digitally, resolution can be measured and precisely manipulated, and low resolution imaging also has its uses. Degraded low resolution images are used for cataloging purposes in the Kodak Picture Exchange as a visual form of general description of the product; the low resolution images give some idea of the full resolution ones while remaining unusable (and thus unpiratable). Users pay a fee to search the low-resolution images in the database, and can order high-resolution copies. And, in police or courtroom videos shown on television, low resolution is sometimes used in areas of the image to obscure people's identities.

Temporal resolution, or frame rate, is a measure of the number of images taken over a period of time (for example, most film is shot at 24 frames per second). The effects of temporal resolution are most noticeable when transfers are made between media. For example, the change from 24 frames per second (film) to 60 interlaced fields per second (television) in a telecine transfer results in a slight jerkiness of motion, since certain frames appear longer than others. Temporal resolution reductions are also evident in the stuttering jumpiness of compressed *QuickTime* movie clips, and particularly noticeable in slow-scan amateur television which runs at only a few frames a second.

Color resolution is the measure of the number of different levels of tone, hue, and saturation that can be depicted in the image; in a black and white image, this amounts to a gray scale. In analog media there are upper and lower limits to the dynamic range of color reproduction, but without discrete divisions in between them, as there are in digital. Thus, in all digital images, posterization occurs to some degree.

Once the image has been recorded, it exists and is stored as a signal, increasing the flexibility of its use as a document. Photojournalists with digital cameras can send their images immediately to an editor, signatures can be sent by fax, and the Voyager spacecraft can return images of Jupiter and its moons. Archives can keep images preserved intact in a form almost immune to physical decay, and provide greater access to stored images.

On the other hand, the ease of creating perfect copies without generational loss may make some archives wary about releasing materials to the public in digital form. Stored as a signal, the image is dependent upon interpretation to be correctly transduced back into its visual form. This fluidity means that even a carefully recorded image may become distorted when output by different means, even when none of the damage is intentional. A dramatic example of such differences occurred during the week of October 18, 1993. That week, *Time* magazine, *U. S. News & World Report*, and *Newsweek* all carried a picture of captured U. S. Army Chief Warrant Officer Michael Durant on their front covers. The photos were all close-ups taken from CNN videotape footage. The reproduction of the photograph, however, differed markedly from one cover to another. Even if one ignores differences in presentation, framing, and surrounding text captions, it is obvious even at a casual glance that the versions of the image vary widely; and as a close-up of a wounded man held hostage during a war, the image has an emotional impact which is inflected by discrepancies in image reproduction. Taken from video, the images are all slightly blurred and fuzzy; but even here there are differences.

The image on *Newsweek*'s cover is blown up the largest, Durant's face filling the page, and the horizontal lines making up the image are very noticeable, as every vertical edge has a slight zigzagging alias. The color is the most washed out of the three covers, and some posterization is apparent. Parts of the image are almost washed out in fogginess. The colors are nearly all browns, with little saturation. Durant's skin is a pale yellow-brown, giving him a slightly sickly appearance, and the cuts on his face are darker brown, and fading in the haziness. Due to the vertical aliasing, the image appears to be clearly from video, and the disruptive effect it has on the image is slightly distancing, along with the lack of color saturation and narrow tonal range.

The image on *U. S. News & World Report*'s cover is soft and no aliasing is apparent. The tonal range is greater, and Durant's hair and eyebrows, which had appeared brown on *Newsweek*'s cover, appear as black here. The color is not as faded, and Durant's skin color is pink and closer to a healthier appearance. There is less of a catchlight in his eyes, and what remains is softer, giving him a calmer appearance than on the *Newsweek* cover, where a stronger, more specular catchlight makes him appear alarmed. Likewise, the cuts on his face also appear as black, and are indistinguishable from mud or dirt; the overall effect is of a much healthier and calmer Durant.

The image on *Time*'s cover appears to be slightly compressed horizontally (possibly to accommodate the text caption to the left of the image). This distortion does not appear on the other two covers, and the horizontal compression makes Durant's head appear thinner and slightly more gaunt. The image is also soft, but there is more color saturation. Durant's face has a yellowish cast, making him look even more ill than on the *Newsweek* cover. The cuts on his face are red, and are clearly recent injuries, not mud or dirt; the Durant of the *Time* cover appears the most ill of the three (although the whites of his eyes are brighter). Curiously, the *Time* cover also has completely replaced the background behind Durant. While the other two covers feature the background from the videotape —Durant is in front of an indistinct white wall— *Time* has replaced the wall with a high-resolution background of a blue sky with clouds!

While there is no way to determine which image of Durant is closest to the original CNN videotape, the changed background and stretched image seem to indicate the *Time* cover as the most manipulated, yet the redness of the cuts and Durant's ghastly pallor may best convey the officer's pain which all three covers attempt to express. But even though the *Newsweek and U. S. News & World Report* covers are more consistent, they still vary a great deal from each other, and there is no way of telling what the original CNN footage looked like (much less what Durant himself actually looked like).

The fluidity of the digital image, and its quantization into a set of numerical values mapped onto a grid, mean that in addition to varying color renditions, digital images can be mathematically scaled to fit other resolutions; that is, an image can be reproduced at different resolutions with varying degrees of detail. This scalability makes it possible to set image standards and to translate images from one medium into another, allowing greater access for potential viewers and greater exposure for the image, although with far less control of the context in which the image appears. An image may be turned into various forms of hard copy, or be rescaled again to fit other media. Such scaling is usually not seen as a manipulative act, because it is performed algorithmically, and because scaling itself is a common visual experience in nature; detail decreases with distance, color saturation is lost due to atmospheric effects and the averaging of adjacent colors, and persistence of vision limits the eye's temporal resolution. Scaling is also often unavoidable when translating an image between media, and is thus often present even when unwanted.

Scaling has always been available in analog media, but digital media introduce ways to compare and measure the information value of images precisely; resolution is measured by the size of the grid being used. In the past, with less developed computer technology, the price paid for this kind of control was lower resolution images, which were still (then) prohibitively expensive and extremely time-consuming to work with. And as digital imaging improved, so did filmstocks, keeping their lead in image quality. But since the appearance of Eastman Kodak's Cineon digital film system in 1992, digital film has been able to more closely compete with the resolving power of filmstock. With lasers that can scan and print images onto film at 166 lines per millimeter, a full resolution image (2664 lines x 3656 pixels) appears visually identical to camera negative, and half resolution (1332 lines x 1828 pixels) and even one fourth resolution (666 lines x 914 pixels) are superior to video.

Less expensive technology is also improving, and even desktop systems are increasingly able to perform high-level image manipulation. Lower costs and compatibility between digital media may induce film and video makers to use and combine images from different media, even as they try to maintain the image's indexical and documentary values. Traditionally, discussions of documentary truth value have centered around content and formal aspects like shooting style, editing, and sound, instead of issues of resolution, medium, or scalability. But regardless of intent, even algorithmic processes like quantization and scaling have certain effects upon the documentary value of an image, causing changes in the *pictorial*, *semiotic*, *commercial*, and *psychological* value of the image.

First and foremost, the image is often a picture. Pictorially, a drop in resolution during scaling means a loss of detail, similar to a softening of focus. Since each pixel's value is generated by averaging the light rays falling on that part of the picture plane that the pixel covers, lower resolution images have fewer pixels and less information. The underlying quantizing grid will be more noticeable, calling attention to form at the expense of content, making the picture less transparent. Pictorial value of an image is dependent on its spatial, temporal and color resolution, and a drop in resolution results in less information and detail, less Z-axis depth possible, and a greater reliance on the spectator's Gestalt processes. This in turn creates a greater emphasis on the image's form, less transparency and perhaps a wider margin of ambiguity, a greater awareness of the image-viewing process, and less of what Richard Wollheim calls the illusion of "seeing-through", that is, the ability to see the object depicted without consciously attending to the surface of the image.[3] The jerkiness of motion caused by low frame rates also destroys the illusion, calling attention to the stasis of individual images rather than producing an illusion of movement. Finally, low resolution reduces the apparent depth possible in an image because of the lack of detail. Even in stereo images, apparent depth on the Z-axis is dependent on parallax generated by offsets along the X-axis, which are limited by resolution.[4]

Generational loss in an image will decrease pictorial value, but, like low resolution imagery, it also has its uses. In one sense, generational loss is itself an index of a film or video's indexicality; it is the mark of mediation, a measure of the link to the camera original, a trace and effect of the duplicating and transfers that have occurred. Generational loss indicates the chain of reproductions linking the print to the original, just as the dirt and scratches on a particular print indicate the history of the print itself; as copies are made of copies, image quality is lost. Digital remastering to remove scratches, dirt, dust, noise, and so on is the removing of an individual print's history. Generational loss indicates the existence of the original, to which it is being compared. (The loss of generational loss brought about by digital imaging, then, is a loss of one kind of indexicality.)

As an index, the image has a certain indexical value. The image's importance as record or evidence requires a strong link to its referent, and a process like scaling will change this link, altering the image's semiotic value. In the shift from higher to lower resolutions, the light from the object is averaged into fewer and fewer pixels on the picture plane; at the lowest possible resolution of one pixel, the image would become a uniform color field, and practically all the image information would be lost.

As a loss of resolution reduces detail, the image loses specificity; at a high resolution, an image of a tree, for example, is identifiable as a specific tree, but at increasingly lower resolutions, although still identifiable as a tree, it ceases to be identifiable as any *specific* tree, but rather as a generalized concept of a tree, in the same way a pencil sketch might denote "tree" without being a drawing of any specific tree. As resolution decreases, images grow less indexical, and more iconic, as detail decreases and large-scale structure is emphasized. Since the image's evidentiary status and "authenticity" rely on its ability to record specific detail, this status diminishes as resolution decreases.

Besides the drop in resolution, the rate at which an image's specificity is lost depends on a number of other factors. Certain objects remain identifiable in silhouette while others depend on size, coloration, composition, or different levels of detail. For moving images, motion helps viewers distinguish figure from ground. And an image's past history as an icon and its familiarity with viewers can determine how low of a resolution the image can survive and still be recognizable. Salvador Dali's painting, entitled *Gala Contemplating the Mediterranean Sea Which at Twenty Meters Becomes a Portrait of Abraham Lincoln,* uses Gala, the sky and wall tiles surrounding a window as large pixels to create Leon Harmon's famous low resolution digital image of Lincoln based on the familiar Matthew Brady portrait also found on U.S. currency.[5] Seen at a distance, detail is obscured and Lincoln becomes more visible; a hint of this is provided by Dali in miniature in the lower left hand corner of the painting. Lincoln survives this drop in resolution primarily because this particular image of him has already become an icon (literally and figuratively) in American culture.

The iconic nature of low-resolution imagery can serve to provoke the viewer's imagination. Early home video game systems, like the Atari 2600, had very limited memories and simple, blocky, low-resolution graphics. These simple iconic figures left practically everything to the player's imagination. For example, the dragons in Warren Robinett's *Adventure* for the 2600 were so blocky that their features were barely discernible; their jaws for example, were horizontal rows of blocks when closed, and two diagonals of three blocks each (one going up and one going down) when open. "Movement" was simply changing from one to the other. Because the representation was so simple in design, the player had to imagine almost everything, and what was imagined could vary from person to person, or even from game to game, the same way that a description of something in a novel will vary from person to person or from one reading to the next. By contrast, higher resolution games have less of this sort of flexibility, and their authors have more control over how the graphics appear, leaving less to the player's imagination.

Digital scaling, then, makes indexicality a matter of degree, blurring the distinction between index and icon. On the other side of the Peircean triad, digital imaging blurs the distinction between index and symbol, often pulling the image toward the symbolic. Unlike some computer graphics, indexical images are in some way linked to a referent in the outside world which produces a field or effect that the image-capture device can record. Photographs use light rays, radar uses radio waves, sonar and ultrasound use sound, cloud chambers detect particle emissions, and scanning tunneling microscopes generate topographic images of atomic structures through precise measurements of surface charge. Recorded data is then displayed as an array of visible light values, which correspond to certain properties, resulting in an image requiring some prior knowledge in order to be "read" correctly; an explanation is needed to indicate the way in which the image is linked to the referent. In each case information read from a referent is used to form an image, but the relationship is not always one of visual resemblance. Often some foreknowledge of the image-capture process, and the way in enacts a link to the referent, is needed in order to decipher what exactly is being represented and to explain how the image is related to the phenomenon in question, as the indexical value is not always clear. Objects may be falsely colored, represented by geometric solids, or turned into an abstract display of information arbitrarily connected to its referent —and thus almost completely symbolic. This move towards abstraction and assigned meaning is what makes the image more symbolic, as the indexical link to the referent becomes less clear and more arbitrarily defined.

Digital imagery's move away from the indexical and toward the iconic and symbolic makes it more akin to spoken language, another digital code both symbolic and iconic, if we include onomatopoeic words that bear a resemblance to the sounds they represent. Similar to language, the image's gain in power and flexibility is offset by potential ambiguity and a weakened or abstracted link to the referent. It becomes more conceptual, more like memory and less like direct experience, in the same way a low-resolution image compares to a hazy recollection.

Just as resolution can play a role in determining an image's pictorial and semiotic value, it can also affect an image's commercial value (as we have seen with low-resolution images used for cataloging purposes). In order to protect their market value, images are often copyrighted. Obviously, higher resolution images are more unique and copyrightable, but at what resolution does a picture *become* copyrightable? No one can copyright a single pixel, or 2 x 2 grid, or 3 x 3, or 4 x 4... but at what size can a copyright be obtained? Corporate logos are copyrightable, and are often simple in design, but how simple can one be without entering public domain?

Copyright laws protect against paraphrasing from a copyrighted work; according to copyright law the making of charcoal sketches of frames of a copyrighted film may be considered as an infringement.[6] Derivative works made from a copyrighted work are also protected. Low-resolution images, which are scaled-down versions of higher resolution ones, can be considered paraphrases or even derivative works, since they can differ greatly from the original high resolution images, to the point where one cannot really say the two images are the same. And if the case for infringement hinges on the recognizability of the appropriated work, who determines at what resolution the image becomes recognizable, and thus an infringement?

Besides issues of copyright law, there is another direct correspondence between the pictorial value and commercial value of an image. As discussed briefly in chapter four, higher resolution images require more computer memory and processing time than lower resolution images, and thus cost more. Thus, as computer memory is the currency of the digital image industry, it is a useful way of comparing images regardless of content. For example, in 125.7 gigabytes of memory, you could store one minute of IMAX film, or slightly over two and a half minutes of 35mm film, or a little over 52 minutes of video.[7] This flexibility is even built into some consumer products; Ricoh Co. Ltd's DC-1 digital camera can record up to 492 still images, 100 minutes of sound, or four video scenes of five seconds each.

Budgetary constraints can force decisions between high quality images and a high quantity of images, even within the same medium. Resolution loss is often seen as unimportant, and gladly traded for gains like accessibility, temporal length, cost, compatibility, availability, and manipulative potential, especially in the area of home use. Since machines displaying lower-resolution imagery are usually cheaper and more readily available, accessibility can be a factor; compare the cost and convenience of going out to a movie to staying at home and renting a video. Within video, the debate between letterboxing and pan-and-scan can be seen in terms of resolution allocation; because the 1.33:1 aspect ratio of television does not fit most film frames, two options —letterboxing and pan-and-scan— are available. While both are reductions in resolution, pan-and-scan crops the image and fills the video screen, whereas letterboxing displays the entire image, although at a lower resolution than pan-and-scan does, since it does not use the whole screen.

Quality versus quantity is apparent in the home use of different formats; video is used in terms of hours, super-8 movie film was used in terms of minutes, and 35mm is reserved for single images; the greater the number of images produced in a medium, the lower resolution of any individual image. Now, however, companies are encouraging consumers to use video even for their still images. Systems like Canon's Xapshot and Kodak's PhotoCD market their systems based on their ability to archive photos, and do not mention the great drop in resolution and image quality when 35mm film images are transferred to video. Ironically enough, one of the advantages of these systems often cited in magazine ads, is image size; although the resolution is much lower, the display area —the television set— is supposedly better because the screen size is much larger than a photographic print ("Panoramic" cameras are also used to produce images which are larger in size but lower in resolution). What they are essentially saying in their advertisements is that although they have far fewer pixels, the pixels are much bigger.

Although a TV screen may be bigger in area, the photograph has a *much* higher resolution than video; and besides that, the photograph must be cropped in order to fit into the television aspect ratio. But of course the ad does not mention this cropping and dropping of resolution (even though it can be seen in the photograph within the ad), or the fact that photo albums are more mobile, do not require electricity or television sets, and cost only a fraction of the machinery otherwise needed to view photos.

Accessibility as a tradeoff for resolution does have its commercial potential, and can sometimes seem like a good exchange. An example is the Voyager Company's laserdisc series of the Louvre, a three-volume set featuring more than 5,000 works of art in over 35,000 images, many of which are close-ups of details of the works which try to overcome the resolution loss, by fragmenting works and reframing them. The discs increase the accessibility of the artworks, but at $100 a set (not to mention the cost of a laserdisc player) they aren't increasing it to too wide of an audience. Books, of course, may be able to do the job better, but they cannot display 5,000 works of art with reasonable resolution either; quality versus quantity once again. Some of these problems may be alleviated with larger computer memory capacities, although display devices would also have to be changed. And no matter how good technology gets, the high end will always be higher than the low end; industrial forms of entertainment will have to be better than home formats in order to compete with them, and adaptation of these forms into home versions will always be a compromise.

Two other examples of accessibility and marketability over resolution are Apple's *QuickTime* video and Fisher-Price's PXL2000 video camera for kids. With *QuickTime*, video images can be imported into a computer and displayed and run at various resolutions. Users can set the compression rate, color depth, frame rate, and scaling of the image in order to fit into any given amount of memory. This manipulative potential exists at the expense of resolution, but it does allow the user to experiment with the variables of the stored image, and by forcing the user to address issues of memory allocation, may even enhance awareness of the tradeoffs occurring.

Although *QuickTime* videos potentially have some of the lowest resolution imagery, Fisher-Price's PXL2000, sometimes referred to as Pixelvision, is likely the medium with the lowest resolution, with only

2000 pixels to an image. Appearing in 1987, this small, simple-to-operate video camera for children ran on double-A batteries and could fit 11 minutes of its grainy black and white images on a 90 minute audio cassette tape. Priced at $99, the camera did not sell and was quickly discontinued, although it found popularity with avant-garde video artists who even held PXL2000 video festivals in Los Angeles and Baltimore in 1991. One of the participants, Spike Stewart, was even working on a feature-length adaptation of Shakespeare's *Twelfth Night* shot on the PXL2000.[8]

From the PXL2000 video at 2000 pixels per frame to digitized IMAX film frames at 22,880,000 pixels per frame, digitization, by quantizing resolution into a measurable quantity, allows a measure of an image's indexicality to be made, as each pixel represents a sample of light taken from the location where the image was captured. Resolution takes up computer memory and rendering time, both of which cost money. Through the technology involved in digital imaging, then, the image's link to its referent can vary and depends on what the customer can afford. Through quantization and digitization, indexicality has become a commodity.

Digital imaging and processing technology continue to evolve. Fractal compression techniques developed by Iterated Systems, allowing compression ratios as high as 20,000 to 1, may even eliminate resolution as a reliable measure of indexicality. Fractal compression schemes store images as equations; the image is broken up into a series of tiles and then each tile is mathematically encoded as a transformation of other areas in the image. The high compression rates greatly reduce the memory needed to store images, and because the image is no longer stored with any inherent size, you can decompress it at any resolution you like, as opposed to decompressing the original and then scaling it to fit. Thus, the image is resolution independent, and it may no longer be clear what the original resolution of the image was.[9] According to one article;

> Iterated also has stumbled on another revolutionary aspect of the technology called "fractal image enhancement" - a process than can actually add details missing from the uncompressed scanned image or digital file. The process works by calculating what information was probably left out of the image when it was originally broken down into a grid of pixels. This technique could also allow images to be greatly enlarged without showing pixel chunks or otherwise losing detail...[10]

Not only will more money get you more indexicality, but this technique claims to give you more indexicality than the original full-resolution image had at the time of its capture!

Actually, the "added detail" is created through the interpolation of pixel values, bridging the gap between pixels with a smooth transition. Traditionally, resolution has played an important role in determined what information does —and does not— exist within an image. In films like *Blow-Up* (1966), *Blade Runner* (1982), or *Patriot Games* (1992), images are blown up and their minute details searched and enhanced or even explored, sometimes far beyond what can be done with real images. However, in real-life counterparts to these image searches, from government and military uses of image enhancement to the videotape of the Rodney King beating which was heavily scrutinized in the courtroom, resolution defines the limits of the informational value of the image. Resolution-independent imagery threatens to blur the boundaries not only between pixels, but between information recorded at the image-capture stage and mere mathematical speculation. The high rates of compression possible with fractal techniques means that it will likely play a large role in the image compression industry, and particularly in the areas of film and video. Even if both compressed and uncompressed versions of images are made available to buyers, it seems likely that the far more economical compressed ones will win out.

If indexicality is seen as a commodity subject to forces of commerce, it becomes necessary to ask what value it has to the spectator/consumer of the digital image. The importance of indexicality to documentary may be obvious from a theoretical standpoint, but to what degree is it valued by the public? The increased mutability and manipulative potential available in digital imaging certainly engenders more skepticism, as the proliferation of special effects films displays digital imaging's capabilities. Faked photographs and images used out of context either by the media, advertisers, or the government may diminish photography's ontological status in the public's eye somewhat, and plenty of home computer programs, such as Adobe PhotoShop, have image-altering capabilities that could potentially demystify the image-altering process. Since images are often presented as evidence rather than as statements requiring proof, to what extent does a distrust of images spill over into other areas? Or if people have no way of knowing one from another, might they cease to care?

When growing disbelief is combined along with other factors, indexicality may not appear to be as important to people as it once was. Lower-resolution media are generally cheaper and make ownership, access, and manipulative control possible on the home computer, and these factors are often valued over resolution in the commercial marketplace, where such tradeoffs are promoted. Movies available on videotape are cheaper and more controllable than theater screenings, and low resolution *QuickTime* clips are more manipulatable, faster, and take less memory than full resolution video.

On the other hand, indexicality can sometimes be "in the eye of the beholder", and *not* very dependent on resolution. Certain contexts and political situations can construct a high degree of belief of an image's representation of reality, even when the image is muddy or at a low resolution. The sharpness or clarity of the image was certainly not a factor for most people in George Holliday's video of the Rodney King beating; it managed to polarize most people. Shot on video at a distance and at night, the sometimes-murky low-resolution image left some ambiguity as to what hits of the policemen's clubs connected, and became a subject long debated in court, as the video imagery was repeatedly examined. The image's ambiguity led to long deliberation, an acquittal, and rioting; although it is difficult to say, owing to the politics involved, one can speculate that had the image been at a higher resolution, the court case might have been more clear cut and more quickly resolved. Higher resolution might even have prevented the riots. In any event, it is clear that other factors can fill in where resolution does not, and they *will* fill in, when resolution is lacking and ambiguity is high.

An image's indexicality is only as strong as its weakest link to its referent. The ethereality, random access, manipulative potential, and varying resolutions of digital images make them akin to spoken language, just as a hazy videotape image is more like remembering than actual experience. Since images are usually read by and designed to have an effect on a viewing subject, changes in an image's pictorial, semiotic, and commercial values alter an image's psychological value. As the form and use of the image play a role in its transparency and the affect it elicits, they also help determine the degree of involvement or distanciation between image and viewer. There is the danger that images will come to seem like data to be read and analyzed, objects so completely constructed, controlled, and manipulated, that they fail to evoke any response apart from passive consumption and detached intellectual analysis. Even if an audience is provided with proof that no alterations were made, they may remain unmoved, seeing all images as merely entertainment, ideology, or both. In the academic realm, the analytical viewer concerned with formal elements and deconstructive activities risks distanciation and an inability to empathize with the people represented in the imagery; when ideological manipulation is expected, informed intellectual viewing practices are valorized while emotional response becomes suspect. Images may come to be thought of as little more than illustrations of ideas, speculations, and artistic constructions, and much of their power of documentation will be lost.

Traditionally, discussions of indexical value have centered around content and formal aspects such as shooting, editing, and sound, while ignoring issues of resolution, medium, and scalability, but the assurance of indexicality in photography was always an illusion. From early on, composite photographs were made, and photographs were misleadingly labeled, such as Hippolyte Bayard's 1840 "Self Portrait as a Drowned Man".[11] In one of the very first photographic experiments, Joseph Nicéphore Niépce's 1826 *View From His Window at Le Gras*, a pewter plate exposure which took 8 hours, sunlight can be seen illuminating buildings on both sides of Niépce's courtyard, a result of the long exposure time.[12] Though of course not a deliberate deception, the picture depicts impossible lighting conditions, and some description of how it was made is necessary to explain it. As acknowledgment of digital photography's mutability and potential for deception comes to light, it reveals issues which apply to traditional photography and has helped to reveal the potential covert deception that has always been possible in analog film-based photography.

The image has not lost all of its indexicality and documentary value, but indexicality has become more difficult to recognize when it is present, since it can be present to different degrees and in different kinds of linkages to the referent. With today's emphasis on access and control, indexicality may simply no longer have the same importance to society that it once did. Ironically, this reduction of the value of indexicality occurs at the same time as its commodification (or perhaps this reduction has occurred *because* of its commodification). The condition of postmodernism, that Frederic Jameson describes as the "cultural dominant" of the late twentieth century, can also be seen as reflecting these trends. The ideas behind postmodernism —the reappropriation of past styles, the effacing of history, the waning of affect, the reconfiguring of connotation— all bear witness to the loss of importance of indexicality.

In asking whether the value of indexicality has changed, it should also be asked whether indexicality itself has changed. Digital technology does not necessarily destroy the "authenticity" of the image and "its power to lay bare the realities" that Bazin admired.[13] Instead, it challenges the notion of authenticity and indexicality by making them *matters of degree*, placing them on a sliding scale in which the resolution of the image determines the strength of the link to the referent, while the algorithms and hardware of the image-capture apparatus determine the nature of the link. Technologies like fractal image compression also redefine indexicality through resolution-independent imagery, which may require original resolutions to be recorded along with image files.

The digital image also demonstrates the continuum between representation and abstraction; while a high resolution image can appear photorealistic, close-ups of portions of it reveal it to be a field of solid color square pixels. A single pixel is a color field, an abstraction which only gains meaning in context among many other such pixels, involving a "double articulation" similar to that of phonemes in speech. Since they are indivisible, pixels are the units of physical indexicality of the image, describing properties of an object's appearance, while at the same time, as color fields, they are abstractions. Aside from color, then, their indexical nature lies in the way they are combined with other pixels to form the image; it is this interaction of parts which is similar to conceptual forms of indexicality.

Digital images can be made in which the objects depicted have no physical referents; in such a case one could claim that the 'referent' is a mathematical model in the form of a data set, although this is ontologically a far cry from what we think of as a "real" object. Likewise, computer simulations of events can have a similar status. But what happens when the "data set" for a simulation is taken from the physical world? The data set can become an indexical link to an existing referent, like a landscape which is photographed and used as a model in a flight simulator. The digital image, then, can have two types of indexicality; a *perceptual* indexicality, which is similar to the that of the analog image and deals with *visual appearance*, and a kind of *conceptual* indexicality as is present in the data sets represented in the digital imagery of computer simulations, which represent *behavior* more than appearance. Both of these types of indexical links are found in computer imaging and simulation, which may be considered as a subjunctive form of documentary.

Subjunctive Documentary
Computer Imaging and Simulation

While most documentaries are concerned with documenting events that have happened in the past, and attempt to make photographic records of them, computer imaging and simulation are concerned with what *could be*, *would be* or *might have been*; they form a subgenre of documentary we might call *subjunctive* documentary, following the use of *subjunctive* as a grammatical tense. At first glance such a term might appear to be an oxymoron, but in some ways, there is no more contradiction here than in any other form of documentary. The last of the above conditionals, *what might have been*, applies to all documentary film and video, since all are subjective and incomplete, reconstructing events to varying degrees through existing objects, documents, and personal recollections. However, by translating invisible entities (those beyond the range of human vision) or mathematical ideas into visible analogs, computer simulation has allowed the *conceptual* world to enter the *perceptual* one. It has created new ways in which an image can be linked to an actual object or event, with mathematically-reconstructed simulacra used as representations, standing in for photographic images. By narrowing and elongating the indexical link and combining it with extrapolation or speculation, computer imaging and simulation may suggest that there is a difference between "documentary film" and "non-fiction film", especially when one is documenting the subjunctive.

As discussed in the last section, many people are willing to trade some (though perhaps not all) indexical linkage for new knowledge which would otherwise be unattainable with the stricter requirements of indexical linkage that were once needed to empirically validate knowledge. Many of these requirements have to do with observation, the visual verification of one's data. In *Techniques of the Observer*, Jonathan Crary describes transformations in the notion of visuality that occurred in the first half of the nineteenth century. Writing about the relationship between the eye and the optical apparatus, he states;

> During the seventeenth and eighteenth centuries that relationship had been essentially metaphoric; the eye and the camera obscura or the eye and the telescope or microscope were allied by a conceptual similarity, in which the authority of an ideal eye remained unchallenged. Beginning in the nineteenth century, the relation between eye and apparatus becomes one of metonomy: both were now contiguous instruments on the same plane of operation, with varying capabilities and features. The limits and deficiencies of one will be complemented by the other, and vice versa.[14]

The notion that observation could be performed through the mediation of instruments did not find immediate acceptance; for example, Auguste Comte, the founder of positivism, deprecated the microscope, which to him represented the attempt to go beyond direct observation. In *Instrumental Realism*, Don Ihde writes that seventeenth century Aristotelians objected to Galileo's telescope, and Xavier Bichat, the founder of histology (the study of living tissues) distrusted the microscope and would not allow them into his laboratory.[15] Idhe points out, however, that some of the objection was due to imperfect technology which produced blurred images. In one sense, a blurred image is an indication that the technology has reached its limits; sharp images with dubious indexical linkages may be more harmful in that their shortcomings are less noticeable. The instrument does not indicate that which cannot be seen through it, and so this must be taken into account when studying objects viewed with it.

The use of imaging technologies is essential to much of modern science, but the issues regarding the status of the entities studied through them are still a matter of debate. In his discussion of instrumental realism, Ihde describes the work of philosopher Patrick Heelan;

> Heelan's position regarding instrumentation is clearly an enthusiastic endorsement of a "seeing" with instruments, albeit cast in the hermeneutic terms of "reading" a "text" provided by instruments. Nevertheless, this "reading" is held to have all the qualities of a perceptual "seeing". Thus, one can say that Heelan's is a specialized but liberal interpretation of "seeing with" an instrument.[16]

Ihde gives Heelan's example of a measured perception which is both hermeneutical and perceptual; that of reading a thermometer to learn the temperature, an act which requires no knowledge of thermodynamics, but rather, an understanding of how the instrument is read. Ihde points out that another philosopher, Robert Ackerman, has a more cautious position;

> On the spectrum of a reality-status for instrumentally delivered entities, Ackerman takes the most conservative position. He argues both for the largest degree of ambiguity relating to what he calls "data domains" which are text-like, and for considerable skepticism relating to the (hermeneutic) interpretative process. "Instrumental means only produce a data text whose relationship to nature is problematic." And, "the features of the world revealed to experiment cannot be philosophically proven to be revealing of the world's properties." (I am not sure whether Ackerman holds any counterpart thesis that other means of analyzing world properties can be philosophically proven!) Yet, examination of the interpretive result relating to such data domains turns out to be the same as for other knowledge claims.[17]

Thus, along with optical instruments like the microscope and telescope, we could include the camera and the record it produces. Photographic technology and processes, since their invention, have come to be valued as records and evidence in the areas of science, law, and medicine.[18] Most people's reliance on media for knowledge of the world is another instance of belief in the recorded image.

Photography, as a source of instrumental realism, differed from prior instruments because of its ability to store an image. The viewing of entities using telescopes, microscopes, and other optical apparatuses, occurred in real time, with an unbroken link between subject and observer, which were both physically present on either end of the link. The camera, as a system of lenses, was similar to the telescope and microscope, but it had the ability to store the images seen, saving them for later examination, even after the subject was gone. The photograph's image was not "live" like the images viewed in other optical instruments; but its fixity provided a means of making a record, which could be analyzed later, like a scientific journal. The photograph bridged the gap between *observation* and *documentation*.

The documentary value accorded the photograph was due in part to the acceptance of other optical apparatuses as epistemologically sound instruments, and could be seen as an extension of them (other imaging technologies, such as video, are accorded similar status). As the computer came to be trusted as a scientific tool, it was only a matter of time that it, too, would be combined with the imaging apparatus, especially in its revolutionization of the idea of the retrievable document. The concept of the stored image is analogous to the stored program, and the digital image is the marriage of the two.

Computer imaging is often indexically less direct than film-based photography, due to the active mediation of hardware and software, as well as the storage of the image as a signal instead of a fixed record. Like the magnification produced by the microscope or telescope, computer imaging is often an extension of the camera into realms indirectly available to human experience. One of the functions of the *Voyager* spacecraft, for example, was to send back images of the outer planets of the solar system. Images of the planets taken by its on-board camera were transmitted electronically back to earth, composited, and arranged in sequence to produce moving imagery, which contained far more detail than what an earthbound observer could produce. Physical distance is not the only boundary overcome; many forms of computer imaging record light waves or energies that fall outside the spectrum of visible light —infrared, ultraviolet, radio waves, and so on— and transduce them into the visible portion of the spectrum, creating visual imagery from recorded data. Likewise, forms of medical imaging such as computerized axial tomography (CAT) scans, positron emission tomography (PET) scans, and magnetic resonance imaging (MRI), use radio waves (and ultrasound which uses sound waves) to construct an image or 3-D model, in which different tones or colors represent different intensities.

The rendering of these transduced waves into the visible spectrum means tones or colors must be assigned to various frequencies to make them visible. The false coloring making up the image, however, is a step into the subjunctive, since the image is not a record of how the subject appears to the observer, but rather, how it *might* appear, if such frequencies were substituted for frequencies (light waves) within the visible spectrum. Thus it is the relative differences between frequencies which are being documented, not the frequencies themselves. Also, false colors in computer imaging are usually assigned to make the data being displayed more clear, representing another level of mediation between subject and observer.[19]

In some ways, the false coloring found in computer imaging is only one step removed from black and white photography, which shows us things not as they appear to us (in color) but rather as a map of tonal intensities. In this sense, black and white photography could also be considered subjunctive, showing us what things would look like if our retinas contained only rods and no cones, making us sensitive to tonality but not to hue and saturation. Ironically, the emphasis on computer imaging processes can lead to neglect of the real, visible colors which exist in the universe. In "A Universe of Color", David F. Malin describes how the visible spectrum is often overlooked by astronomers, who often prefer to use charged-coupled devices (CCDs), electronic detectors that collect light far more efficiently than photographs can;

> One of the key advantages [film-based] photography has over electronic imaging is that photographic plates can capture high-resolution, sensitive images over an area of virtually unlimited size, for example, across the entire wide field view of the U. K. Schmidt telescope in Australia. In comparison, the largest CCDs measure only a few centimeters across. Photography therefore offers a superior means for recording images of extended astronomical objects such as nebulae and nearby galaxies.
>
> Furthermore, the photographic layer serves as a compact medium for storing vast amounts of visual information. A single 35.6- by 35.6-centimeter plate from the U. K. Schmidt telescope, which records a patch of sky 6.4 degrees on a side, contains the equivalent of several hundred megabytes of data.
>
> Color images present the information packed into an astronomical photograph in an attractive and intuitively obvious way. Indeed, measuring the colors of nebulae, stars and galaxies is central to understanding their composition and physical state. Yet researchers have been slow to appreciate the value of revealing the color inherent in many astronomical objects. Many astronomers consider color an abstract concept, easy to quantify but rarely rendered correctly on film.[20]

In the computer age, film-based photography now seems like a direct process of observation, compared to computer imaging, quite a far cry from the days when the microscope was considered a step beyond direct observation. At the same time, computer imaging technologies have changed the nature of "observation" and what is considered observable.

Many of the entities or energies imaged by the computer are ones which are either too small, too large, or too fast to be visible to traditional optical instruments or the unaided eye. In the photographic world, microphotography, time-lapse, and high-speed photography serve similar purposes, although to a much lesser degree; but in several cases they can be seen as precursors of computer imaging techniques. The electron microscope, for example, is an extension of the optical microscope and the camera, using a focused beam of electrons rather than light to create an image, while the scanning electron microscope produces an image from scan lines in a manner similar to television images. The scanning tunneling microscope, which uses computer imaging, produces an image by scanning a very small area with a fine, microscopic probe which measures surface charge. The closer the probe is to an atom, the stronger the charge is. The charge on the probe changes during the scan, and a reconstructed image of the surface emerges when different colors are assigned to different charge intensities and displayed. Although the process presents a visual image of the atoms, it is not an image in the traditional photographic sense, since traditional photography is impossible on a subatomic scale. The useful magnification power of the light-based microscope is limited by the wavelength of light in the visible spectrum, which cannot be used to image objects smaller than the waves themselves, any more than you could determine much about the shape of a small porcelain figurine with boxing gloves on your hands. Since traditional photography also is light-based, it suffers the same limitations.

The recording of high speed events in particle physics also evolved from photographic techniques to computer imaging techniques in which 25 billion bits of data are produced in one second. The earliest detectors using for monitoring particle collisions were cloud chambers, bubble chambers, and streamer chambers, which all were similar in principle. In these chambers, charged particles were detected when they moved through a medium leaving a trail of ionized atoms behind them. These trails (composed of water droplets, bubbles, or sparks, depending on the chamber) could then be recorded on film. Photographic records were highly detailed and could be interpreted easily, and during the late 1960s computers were used in the analyses of the photographs, although more complicated images required human assistance. However, according to a 1991 article in *Scientific American*;

> Capturing events on film became impossible when accelerators were developed that produced thousands of particles in a second. To record particles at this rate, physicists designed complicated electronic detectors. Because information was gathered in electronic form, computers became an essential tool for making quick decisions during data collection.
>
> Yet physicists still cannot rely exclusively on computers to analyze the data. Computers that automatically inspect events are limited by the expectations of programmers. Such systems can selectively suppress information or obscure unusual phenomena. Until scientists invent a pattern recognition program that works better than the human brain, it will be necessary to produce images of the most complicated and interesting events so that physicists can scrutinize the data.[21]

Towards the end of the article, the authors state that "The detectors, like all complex mechanisms, have certain quirks." And they add that during the test runs of the detectors, "...the programs occasionally produce images showing inconsistencies between the data and the laws of physics. These inconsistencies can arise because of a malfunction in part of the detector, in the data acquisition system or in the data analysis software."[22] In this instance, the desire for more data and finer detail necessitated a move away from conventional media which proved to be too limited for scientific purposes. Although the data may not be as reliable, there are several thousand times more data to examine (quantity over quality).

Sensor technology is also used on the macro scale, for tracking and imaging planes with radar, submarines with sonar, or weather patterns imaged by satellite. Electronic signals and sensors have been used to track whole flocks of birds, and even human beings, some of whom owe their lives to these sensors.[23] NASA's Mission to Planet Earth (MPTE) involves sending up nineteen earth-orbiting satellites over a fifteen-year period (1998-2013) to collect data on environmental conditions such as the greenhouse effect, desertification, ozone depletion, weather cycles, and other ecological information. One of the largest of these satellites generates a terabyte (1,000,000,000,000 bytes) of data every ten days; about as much data contained in the Library of Congress.[24] The data being imaged are collected from all over the globe, and far too extensive to be represented in any single image or sequence of images; and if the satellites are collecting continuously, the data may be coming in faster than anyone can analyze or make sense of it.

In this string of examples, computer imaging gradually moves further and further away from conventional photographic methods of documenting events as the phenomena being studied slip out of visibility, because they are too far away, too small, too quick, or because they lie outside of the visible spectrum. And unlike infrared, ultraviolet, or even X-ray photography, computer imaging involves more interpretation and often must reconstruct objects or events in order to visualize them. Although often displayed in two-dimensional images, the data collected in medical scanning, astronomy, particle physics, and meteorology represent complex events and structures in three dimensions. These reconstructed events can be rendered and viewed from any angle, and be replayed repeatedly for further analysis, as computer simulations.

Computer-generated three-dimensional representations that change over time are more than just reconstructions of an image, they are reconstructions of a system's behavior. In these reconstructions, still imagery becomes moving imagery, and computer imaging becomes computer simulation; like the photograph as a stored image which can be analyzed in the absence of its subject, a computer simulation made from recorded data allows an event to be analyzed after it has occurred. The data in a computer simulation are a series of measurements combining disparate forms of other data, many of which are nonphotographic. These data are linked to their referents in a variety of ways, recording the values of whatever variables they are programmed to represent.

If computer simulations are documentary, they are subjunctive documentary. Their subjunctive nature lies not only in their flexibility in the imaging of events, but in their staging as well; computer simulations are often made from data taken from the outside world, but not always. Just as the digital image does not always have a real world referent, computer simulation can be used to image real or imaginary constructs, or some combination of the two. Computer simulations go beyond the mere recording of data; they are frequently used to study the behavior of dynamic systems, and become the basis of decisions, predictions, and conclusions about them. As a simulation is constructed, and the data set becomes larger and more comprehensive, its indexical link to the physical world grows stronger until the simulation is thought to be sufficiently representative of some portion or aspect of the physical world. The computer allows not only physical indices like visual resemblance to be used, but conceptual indices as well, like gravity or the laws of physics, which are used to govern simulated events. Like the photograph, computer simulation can combine observation and documentation, and as the embodiment of a theory, it can document what *could be*, *would be*, or *might have been*. In many cases, actual experiments and events are represented as measurements and relationships; these are abstracted into a set of laws governing the phenomena, and these laws become the basis for creating the *potential events* of the simulation. Thus the simulation documents possibilities or probabilities, instead of actualities.

Computer simulation has been used in a wide range of applications; for the visualization of purely mathematical constructs, architectural walkthroughs, job training, product design and testing, and experimental scientific research. The basis for all of these simulations is mathematical reconstruction, in which a simplified version of the events being studied is simulated in the computer's memory. Mathematical visualization can give visual form to purely mathematical constructs such as fractal objects, rotating hypercubes and other higher-dimensional forms.[25] It can also be used to illustrate the hypothetical, for example, to show how relativistic distortions would appear to an extremely high-speed traveler (buildings would shift in color and appear to lean in towards the observer). A simulation can give the hypothetical situation an appearance of feasibility, by virtue of its visualization; such concrete imagery can be useful in swaying belief, especially in a society which tends to rely on the image as evidence. Although mathematical visualizations seek to give concrete visual form to abstract ideas, most visualizations seek to recreate objects and situations which are thought to have existed or that could exist in the material world.

Architectural *walk-throughs* or *fly-throughs* are computer simulations of buildings which the user can move through and view from any angle, under various lighting conditions, and with different styles of decor and furniture. They allow a user to get a sense of what a building and its interior will be like before it is built, to make changes, and to decide on a final design. Architects designing on the computer may work closely with these tools while creating the plans for buildings. The limitations inherent in the simulation program, then, may subtly influence the design of the building, and the way the building is depicted as a three-dimensional space will also influence the choices shaping its design. While it allows the visualization of possibilities, it limits those possibilities by providing the language of speculation used to create their visual analogs.

Architectural simulations can be used to speculate about the past as well as the future. Structures that are partially destroyed, no longer exist, or were designed but never built can be visually (and even acoustically) recreated.[26] The Taisei Corporation, Japan's third largest construction firm, uses computer modeling in the design of large-scale structures, including skyscrapers and power plants. They have also made a series of architectural fly-throughs of ancient cities which they reconstructed for the British Museum. Reconstructed from archaeological data, a fly-through of the Sumerian city of Ur begins with a rotating bird's eye view of the city. The point of view revolves around a ziggurat and over rooftops, and then eventually drops down to street level, moving through an alley, under canopied doors, and finally through the interior of one of the buildings, where pottery and furniture can be seen. The sequence combines a map of the city, architecture of individual buildings, and artifacts such as pottery into one unified illustration. Similarly, computer animation is used to bring a reconstructed city of Tenochtitlan back to life in *500 Nations* (1995), a television series documenting Native American history. Although the virtue of such reconstructions is a more holistic presentation of disparate data, it is difficult to tell from the imagery alone where historical evidence ends and speculation begins; the problem is similar to that faced by paleontologists who reconstruct dinosaur skeletons from an incomplete collection of bone fragments. It would seem that computer simulations are not so different, after all, from the constructed interpretations found in documentary film and video, where editing constructs a world and a point of view; they all are representations of partial reconstructions of the past, and can be considered subjunctive to some degree. No simulation or film can ever provide enough information for a reconstruction of past events free of speculation. There are always gaps filled in by the person who constructs the recreation, as well as those who study it; and these may range from educated guesses to unacknowledged assumptions.

History, however, is not the only thing at stake; the belief that computer simulations can represent reality is often relied upon to the point that people's lives depend on the accuracy of those simulations. For example, the flight simulators used in the training of pilots are deemed close enough to the real experience that pilots can obtain their licenses without ever having left the ground. During an actual flight, weather conditions may produce zero visibility, forcing pilots to fly entirely by their instruments and radar; when this happens, the cockpit completely mediates the experience of the pilots, much like a training simulator would. In such cases, ironically, actual conditions simulate the simulator.

Pilots and their passengers aren't the only ones whose lives might depend on the accuracy of simulations; on the ground, surgeons can now practice life-saving operations without opening animals or cadavers. Gene Bylinsky describes a medical simulator developed by High Techsplanations of Rockville, Maryland;

> V. Rakesh Raju, the developer, shows a visitor how he can strip away the pelvic bone of a digital torso on a workstation screen to reveal the kidneys, the bladder, the urethra, and the prostate in vivid color. He can rotate the torso and even zoom inside to traverse the interior. "We

> just went through the colon," says Raju matter-of-factly after we fly through a pinkish tunnel with uneven walls. Organs can be programmed to reflect various ailments.[27]

The attempt at realism even extends to the purely visual side of the simulation. According to a newsletter from Viewpoint Datalabs;

> For the opening of a new digestive center, Traveling Pictures produced an animation featuring Revo Man, showing the intricacies of the digestive system. Viewpoint's anatomy datasets were implemented to produce a result that was almost too real. "We reached our goal of making the sequence look realistic," says Dave Burton, animation director. "When we showed our clients the initial animation, it made some people ill because of the realism! We had to tone it down a little by using transparent and glassy materials."[28]

Even if the simulations look as real as Burton claims, will these abstractions prepare medical students psychologically for the real thing? Can virtual blood replace real blood, and the cutting of an incision in a human being with a sharp metal tool? No matter how accurately human organs are simulated, there will always be an *experiential* gap between real and virtual.

Medical simulations are most useful for analysis and visualization when the data used to construct them are taken directly from life. Like an X-ray, the data is taken from the patient's body and analyzed on-screen. One simulation, of the treatment of an ocular tumor, was shown at a SIGGraph convention and described some of the methods. The clip had the following voice-over;

> The physical form of the eye for this simulation comes from diagnostic ultrasound scans. The shapes of the eye and tumor are captured as digitized slices by using digitized pulse echo ultrasound. The digital slices are stretched into a rectangular format and used as the basis for reconstructing solid models. A typical section is shown indicating the tumor and retina. Sweeping through the volume gives the doctors a feel for the data integrity. The model eye is constructed from the image slices by hand tracing. As data slices are peeled away, outlines of the tumor and retina are shown being converted into polygonal objects. The shape of the other structures in the eye are obtained from standard database structures and B scan ultrasound.[29]

The simulation described above combines data gathered from a real source (the patient) with data from outside the patient ("from standard database structures"). The patient's body has become a series of interchangeable parts, for which standardized structures can be substituted; an instance of filling in the gaps with default assumptions.

The substitution of a generalized, "standard" computer model for a specific existing object is common throughout computer simulation. The general model acts as a stand-in for all specific physical versions of the model. Many industries use computer simulation for the design of their products, and an increasing number of them are doing product testing with computer simulation as well. Originally, the purpose of product testing was to see how well a product stood up to forces in the physical world; so there is, perhaps, a certain irony in simulating such tests on the computer— today, product testing could be done physically in order to test not the product itself, but rather, to test how well the simulation simulates the product!

Computer simulated product testing is used for everything from simple containers to complex large-scale products like automobiles, where people's lives may be at stake if the product fails. In 1991, "the world's first digital smashup involving two complete automobiles" occurred inside a computer in Germany. Describing it, Gene Bylinsky writes;

> Such simulations are so accurate and economical that engineers at many automakers now conduct most of their crash tests using computers instead of real cars. In one that models a broadside collision of two Opels, the cars appear on the screen as ghostly X-ray images produced by a Cray supercomputer and software from Mecalog of Paris. The crash unfolds in slow motion. The engineers can freeze the action at any point and study the effect of the impact on the bodywork, the key internal parts, and the dummies inside. Such tests yield detailed results at a cost of about $5,000 per crash—vs. roughly $1 million using real cars.[30]

The cars, crashes, and human beings are all simulated by the computer, and now even the test driving of cars under various conditions is being simulated on the computer;

> Since Renault can't test a new model under real conditions without the risk of leaking its designs to its competitors, it has created a digital testbed for prototyping the aesthetics, road-holding, and drivability of new vehicles.
>
> The first step is to build a database of virtual driving routes. Then designers mock up a complete virtual car with digitized features: weight, size, shock absorption, steering-lock coefficient, and tire profiles. Using a simulator program, the design team then drives the prototype through the database to find out how it behaves.[31]

The environment is simulated along with the product, and both are believed to accurately represent their behavior in the physical world; the product test itself, then, completely relies on how well another product —the computer simulation— works.

Computer simulations and reconstructions are depended upon not only in industry, but in experimental science and law as well, where they routinely are given a status which is almost that of real events. Chapter eight has already described the "Battle of 73 Easting", a digital reconstruction of a Gulf War battle recorded and studied by the military. In the courtroom, computer simulations have been admitted as substantive evidence with independent probative value, that is, they may be used as evidence or proof. In 1985, computer simulations were pivotal in the case concerning the crash of Delta flight 191, whose 14-month trial was the longest in aviation history and in which over 150 million dollars in claims were at stake.[32] Lawyers from both sides made extensive use of computer simulation to support their case, and they were able to establish a precise chronology of events and replay them from multiple perspectives. Most crucial were the films made from the digital on-board flight recorder which survived the crash. These "black boxes" record flight path and instruments, audio of pilots' voices, and airborne radar imagery, and newer versions can log as many as 700 different variables, including the positions of 500 switches. When this data is combined with a map of the terrain, simulations can be made from whatever point of view is desired.[33]

Simulations used in forensics to reconstruct events are sometimes even used to reconstruct the crime itself, resulting in a new, bizarre form of "evidence". Since 1992, Alexander Jason, an expert witness in ballistic events, has been producing computer-generated reconstructions which are used as evidence during criminal proceedings. Ralph Rugoff describes one Jason's productions, made for the murder trial of Jim Mitchell;

> The Mitchell video, which reconstructs the movements of a homicide victim during a sequence of eight gunshots, looks like a primitive video game: the protagonist resembles a faceless robot with orange hair, while bullets appear as bright-red spikes identifying entry and exit wounds. Terse data flash on the screen following each gunshot, but the aura of scientific investigation is occasionally punctured by oddly realistic details—like the way a beer bottle bounces and rolls when it's dropped by the victim. The result is an unnerving viewing experience where a relatively innocent cultural form—the cartoon—is made to serve within a grisly documentary context.
>
> In another Jason animation, produced for a case in Colorado Springs, we see a blue-colored police officer shooting a naked, green woman, who is shown kneeling in bed pointing a silver pistol. The

> woman has breasts—the officer's bullet enters somewhere near her right nipple—but no genitals. In other videos, some figures are depicted with clothing, and in a few cases race is also indicated, yet this realism is tempered by the use of uncanny effects: a torso may abruptly dissolve to reveal the underlying skeletal structure, tracing a bullet's path through the body. In one stunning cinematic flourish, Jason creates a shot from the bullet's point of view, taking us from a policeman's gun, through the victim's chest, and into the wall beyond.[34]

Despite the strange iconic and geometrical appearances of the simulations, people's unquestioning faith in both the documentary quality of the presentations and in the scientific and mathematical means of producing them make it necessary for judges to remind jurors of the unreal and speculatory nature of such simulations. In this sense, some of the abstractions of computer graphics are desirable, since they can act as a reminder of the simulation's construction. They can also be used to distance the jury from the events, in the same way that judges will often allow black and white photographs of a victim's injuries to be used but will not allow color ones because they are felt to unduly influence a jury.

Simulations like Jason's look nothing like the actual crime, but they are believed to represent the events of the crime. Similarly, in experimental science, abstract representations are made to represent events, and in many cases, events that never even happened. In chemistry, computer simulation is used to model such things as molecular bond energies and enzyme catalysis; the objects being modeled are molecules, atoms, and particles, whose individual properties can be satisfactorily defined with a small number of variables, and are visually represented in a variety of ways, one of the more familiar being a three-dimensional array of balls and sticks, similar to the wooden or plastic physical models used before computer modeling. Here, the generalized models used in the simulations come the closest to what they represent, since individual atoms of the same element have no specificity apart from position.

Experiments simulated on a computer come a long way from traditional means of documentation like the photograph or eyewitness reports, and one may ask what indexical connection to the real-world referent remains. It certainly isn't one of visual resemblance, since molecules don't really look like the wiggling balls and sticks. Nor are the simulated reactions reconstructions of specific reactions that actually took place; they are only possible scenarios based on chemical properties, statistics, and the laws of physics. Molecular simulations are based on such things as bond angles, binding energies, and other elementary data, and are used to test and develop theories in a manner similar to physical experiments. In effect, the recreation of dynamics and behavior are their *only* link to the outside world and the only things being documented; the referents are not objects, but laws of physics and descriptions.

In this indexical shift, as mentioned above, actual experiments and observations have been reduced to measurements and relationships, and abstracted into a set of laws; and these laws, in turn, become the basis for the *potential events* seen in the simulation. Theories based on these laws, which fill in incomplete data and understanding of the laws of physics, stretch the indexical link even further. There is a shift from the *perceptual* to the *conceptual*; the image has become an illustration constructed from data, often representing an idea or speculation more than existing objects or actual events.

All simulations are subjunctive to some degree, subjective, and prone to ideological manipulation. Yet, in science and many public sectors, as well as the public's imagination, the mathematical basis behind computer simulation has given it a status similar to (or greater than) photography despite the often much more tenuous indexical linkage. One of the reasons is computer simulation's obscuring of point of view. Documentary theory has shown, in a wide array of writings, how the camera is always subjective, since every image contains an inherent point of view. Simulations, however, do not come with an inherent visual point of view; they can be replayed and rendered from any angle, and reconstructed events can be seen from any point of view desired, including those from the insides of objects. While this omniscience lends an air of objectivity to the displays, it effaces the fact that point-of-view more importantly refers to the programs, theories, and assumptions controlling the simulations, in much the same way that a particular theoretical stance may steer an authorial voice in a photographic documentary. The "point of view" is not visual or perceptual, but conceptual and theoretical, the speculation behind the simulation structuring how everything is seen. As Albert Einstein once remarked, "So far as the laws of mathematics refer to reality, they are not certain. And so far as they are certain, they do not refer to reality."

No simulated "virtual world" can be free of a world-view, and the assumptions behind them are often difficult to discern, given only their visuals. Even barring digital legerdemain and computer simulations deliberately slanted in one direction or another, assumptions and speculations will be shaping the results they produce. Indeed, due to the complexity of the science, mathematics, and program code involved, it may be the case that no single person is aware of all the assumptions involved in a particular simulation. The simulation's subjectivity is a multiple one, layered with the assumptions of several persons and disciplines; the designers, the programmers, the company that sells it, and the field expert who uses it. And, of course, the potential for deliberate manipulations will always remain. It is usually assumed that computer simulations are at least unfolding on a sound mathematical basis, but how can anyone but the programmers be sure?

Computer simulation's speculative nature blurs the line between fiction and non-fiction, and complicates the question of how far an indexical link can be stretched and displaced and still be considered valid in society, as facts get skewed, left out, misinterpreted, or even filled in by theory and speculation. As Jack Weber cautions,

> Precisely because it allows you to see the invisible, the capabilities of visualization, with its pizzazz and drama, can also blind you to what it reveals. To watch the dynamic ebb and flow of market forces or to reach out and touch an enzyme molecule is a seductive experience—so seductive that you can easily forget the approximations and interpolations that went into it.
>
> "One of the problems with data exploration and visualization," says Paul Velleman, president of Data Description (Ithaca, NY), a maker of visualization software, "is that these technologies make it easy to find patterns that may or may not be real." Color, shading, sound, and other dimensions that add realism to visualization are equally capable of making the unreal seem more plausible.[35]

Even the most honest and accurate computer simulations can succumb to software glitches, hardware flaws, and human or computer error, making them unreliable and sometimes potentially dangerous. As in the examples of car testing, medical simulations, court cases, and pilot training, people's lives can depend on computer simulations.[36] And, as Theodore Roszak points out, political leaders often make decisions based on computerized representations and simulations of situations, complete with predicted outcomes, so world affairs may actually depend on software intricacy and accuracy, and the way they develop the user's world-view of the events the simulation supposedly represents.[37]

Computer simulations and systems are enormously long chains who are only as strong as their weakest links. The interconnectedness of computer systems means that errors can propagate to enormous size before they are corrected. Peter Neumann, the moderator of RISKS-forum on the Internet (which is a compendium of reports of disasters due to computer error), describes the collapse of the ARPANET in 1980;

> There was a combination of problems: You had a couple of design flaws, and you had a couple of dropped bits in the hardware. You wound up with a node contaminating all of its neighbors. After a few minutes, every node in the entire network ran out of memory, and it brought the entire network down to its knees. This is a marvelous example because it shows how one simple problem can propagate. That case was very similar to the AT&T collapse of 1990, which had exactly the same mechanism: A bug caused a control signal to propagate that eventually brought down every node in the network repeatedly.[38]

Software glitches can occur due to small errors in source code; in one case, a string of failures at a telephone company were due to three faulty instructions in several million lines of code.[39] Complexity is one of the main reasons for software bugs. "The Risks of Software", an article in *Scientific American*, described the problems of the large program;

> Despite rigorous and systematic testing, most large programs contain some residual bugs when delivered. The reason for this is the complexity of the source code. A program of only a few hundred lines may contain tens of decisions, allowing for thousands of alternative paths of execution (programs for fairly critical applications vary between tens and millions of lines of code). A program can make the wrong decision because the particular inputs that triggered the problem had not been used during the test phase, when defects could be corrected. The situation responsible for such inputs may even have been misunderstood or unanticipated: the designer either "correctly" programmed the wrong reaction or failed to take the situation into account altogether.[40]

According to the authors, the very nature of digital technology itself is partly responsible for the computer's vulnerability;

> The intrinsic behavior of digital systems also hinders the creation of completely reliable software. Many physical systems are fundamentally continuous in that they are described by "well-behaved" functions—that is, very small changes in stimuli produce very small differences in responses. In contrast, the smallest possible perturbation to the state of a digital computer (changing a bit from 0 to 1, for instance), may produce a radical response. A single incorrect character in the specification of a control program for an Atlas rocket, carrying the first U.S. interplanetary spacecraft, Mariner I, ultimately caused the vehicle to veer off course. Both rocket and spacecraft had to be destroyed shortly after launch.[41]

Not only is computer source code ultra-sensitive to error, but the creation of software is not yet even a science;

> Studies have shown that for every six new large-scale software systems that are put into operation, two others are canceled. The average software development project overshoots its schedule by half; larger projects generally do worse. And some three quarters of all large systems are "operating failures" that either do not function as intended or are not used at all.
>
> The art of programming has taken 50 years of continual refinement to reach this stage. By the time it reached 25, the difficulties of building big software loomed so large that in the autumn of 1968 the NATO Science Committee convened some 50 top programmers, computer scientists and captains of industry to plot a course out of what had become known as the software crisis. Although experts could not contrive a road map to guide the industry toward firmer ground, they did coin a name for that distant goal: software engineering, now defined formally as "the application of a systematic, disciplined, quantifiable approach to the development, operation and maintenance of software."
>
> A quarter of a century later software engineering remains a term of aspiration. The vast majority of computer code is still handcrafted from raw programming languages by artisans using techniques they neither measure nor are able to repeat consistently. "It's like musket making was before Eli Whitney," says Brad J. Cox, a professor at George Mason University.[42]

In the computer industry, the design of integrated circuits and microchips has grown so complex that much of it is done using computers. Like so many other industries, testing of the designs is also done on the computer. Computers are being used to simulate computers; thus, if errors occur and go undetected during the design phase, software glitches in today's computers could become hardware flaws in tomorrow's computers. Of course, this doesn't mean the process would be more error-free without a computer; the chips are simply too complex to be manufactured perfectly in any event. For example, Intel Corporation's Pentium microprocessor was given sudden media attention when Thomas Nicely, a mathematics professor in Lynchburg, Virginia, discovered that the chip had a flaw in its floating point divider.[43] The story got a lot of press, some jokes in the computer community (Q: What's another name for the "Intel Inside" sticker they put on Pentiums? A: Warning label), but Intel's promise to replace the chip seemed to quell any fears that might have arisen; the "new" replacement chip, by virtue of its being the "replacement", is seen as being better. But can anyone prove that it is really more error-free than the last chip Intel produced?

At any rate, computer imaging and simulation represent a shift from the *perceptual* to the *conceptual*, a shift that underscores a willingness to exchange direct experience for abstractions that open up of the wide vistas not directly available to the senses. As subjunctive documentary, computer imaging and simulation represent an epistemological shift—from mediation into speculation— as great as the shift from direct knowledge to mediated knowledge.

Spheres of Indexicality

Throughout this book I have used the *spectrum* as an organizational analogy (for example, in looking at various forms of interactivity); as an analogy it is analog rather than digital, a sliding scale rather than a set of discrete levels. Thus, I will now present a gradual, sliding scale of indexicality examining the strength (or length) of indexical linkages, rather than attempt to quantize indexicality into units— the very quantization of indexicality into units would itself be an abstraction of indexicality, and a strain on the linkage that such a concept has to what it represents. I will be suggesting a series of concentric spheres here, (*direct experience*, *mediation*, *speculation*, and *the unknown*), but I want to emphasize that there are no hard and fast boundaries between them; one gradually shifts into the next, as do the colors of the visible spectrum.

In order to chart these spheres of indexical linkages, we must first define a kind of epistemological space, an indexical space surrounding the individual, through which information about the outside world passes to the observer, a space, in between the individual and the intersubjective world, where indexical links exist. The human being, as the endpoint of knowledge about the world, is at the center of these concentric spheres, the recipient of indexical links. The channels through which these links make their way "into" the individual are sensory channels, so the first "sphere" of indexicality, existing in the immediate sensory vicinity, is that of *direct experience*.

Direct experience is everything immediately tangible, audible, visible, and available directly to the senses. Taste, touch, and smell, are perhaps the most direct, as they require physical contact with the object being sensed (smell is the detection of trace amounts of a substance, so it too requires some physical contact). Sight and sound cover a wider area, and are signal based, but the signal is always direct from some physical source which causes it. (This is true for all sights and sounds; even mediated ones must emanate from some physical object near the observer. A sound from a loudspeaker may be coming indirectly from the person speaking over it, but in a physical sense, the sound the observer hears is originating from the mechanism of the loudspeaker itself, and it is this sound to which I refer as direct.) Although most information gained through direct experience refers to the immediate physical vicinity, not all of it does; a person can hear a distant bird call or thunder, for example, or see the sun, feeling its heat and light directly, despite its being millions of miles away.

As soon as something passes out of the direct sensory realm, its indexical linkage weakens slightly, as the object passes into memory. Here direct experience begins to make its gradual transition into the sphere surrounding it, that of *mediation*. Human memory is the most common form of mediation; although it is almost as direct as the senses, memories can fade with time and change with experience, losing the directness they had at the time the impressions were made. But for the most part, human memory is fairly reliable, when one considers how nearly everything in daily life depends on it; the times it fails are small compared to the constant use being made of it. Memories, because they are internal, are so closely tied with the individual that they are nearly inseparable from direct experience, for the whole notion of a consistent intersubjective world existing outside the individual is as dependent on memory as on the senses themselves. Closely associated with the mediation of memory is the mediation of language, which becomes a form of shorthand for representing and expressing concepts and sensory data. Although not a physically-based technology, it is a culturally specific symbolic system imposed on the individual, mediating and shaping thought according to a particular cultural world-view; an example of internalized mediation originating from without.

A slightly stronger degree of mediation (and a further remove) can be found in those devices used to *amplify* the senses (a telescope, microscope, binoculars) or *filter* them (sunglasses or a seashell placed to the ear). Unlike memory and language, these devices are physical objects and entirely external to the user. Although instruments that amplify or filter the senses come between observer and observed, extending the indexical link, the subject being observed is still present at the time of the observation, similar to direct experience. Observer and observed are present together in time, even if separated by the instrument. (Of course there are exceptions; when one looks at stars through a telescope, one sees them as they were many years ago, and not as they are at present. But the same is true of naked eye observations of stars, and it is this similarity to direct observation that I am emphasizing here.)

As mentioned above, the photograph, as a stored image, severed the link between observer and observed in both time and space. Yet while this appeared to be quite a break, the photograph did increase the *image*'s indexical value, and this is precisely the reason Bazin celebrated the photographic image. The photograph changed indexicality in two ways. On one hand, the knowledge gained from looking at a photograph was less than that of looking at the subject of the photograph directly; first, because the subject was absent (or perhaps no longer existed) and second, because it reduced the subject to a monocular, still image, with less detail than the subject. On the other hand, the photograph could be said to increase the image's indexical linkage to its subject, *when compared with all other forms of imaging*; painting, drawing, engraving, and so on depicted their subjects by an artist's hand, in a much more subjective means since it was drawn by hand, while the photograph recorded light more automatically and mechanically, in greater detail than anyone could do by hand, and in a more direct fashion (though of course a photographer is still always present in the process). Thus rather than be seen as less indexical than direct experience, the photograph was seen as more indexical than any other form of imaging, and praised as an advance, its severing of the temporal linkage to the subject seen as a benefit and not a disadvantage.

The photograph, however, was like all other forms of image-making in that the image was made by a person, and more often that not, someone other than the observer of the image. The photograph, then, leaves the sphere of direct experience and fully enters the sphere of mediation. Of course there remains the direct experience of sensing the photograph itself, as an object, but the knowledge gained from the photograph does not come from its own objecthood so much as it does from the image depicted upon its surface. In any form of mediation, the instrument between observer and observed has been replaced by another individual, a mediator; the subject being observed is now experienced directly by the mediator, who then passes along the knowledge to the observer. The link between the mediator and the observer can occur at varying strengths; the mediator may be a spouse or parent, a friend, an acquaintance, a well-known public figure, or a complete stranger. The stronger the link between mediator and observer, the better the observer will be able to judge the reliability of the mediator (and, hopefully, the information conveyed by the mediator). The medium by which the mediator passes on the information can also vary, and effects the indexical linkage; information can be passed on directly through storytelling, conversation, or gesture, or indirectly through the recorded word, written word, photograph, film, video, etc. Thus, in the sphere of mediated information, the link between observer and subject is made up of a number of different links; the direct experience link between the subject and the mediator, the link between the mediator and the medium, and the link between the medium and the observer.

Mediation can occur through a single mediator, or through a series of them. Spheres of indexicality can be nested; the subject is experienced directly by one person, who then becomes a mediator for another person, who then becomes a secondhand mediator for another, and so on. Unless the original information regarding the subject receives an intelligible indexical link to every mediator it passes through, some mark of its passage, there will be no way to tell just how long the link back to the subject really is. An example would be a letter that gets stamped by each post office it passes through, or a quote for which a footnote gives the original source, which is it itself footnoted, along with a note saying where the quote was originally quoted.

Every medium and mediator is imperfect, and none is able to pass on direct experience completely unaltered and in full. A medium, then, whether a mechanical one like film or television, or a human mediator, is a kind of filter for the information passing through it (one of the ideas behind McLuhan's saying, "the medium is the message"). As the number of filters increases, the signal-to-noise ratio will likely increase, with each source either leaving out information, or "filling in" missing information either unintentionally through confabulation or intentionally, through guessing (or through deliberate altering of the information for ulterior motives). As the link between observer and subject grows less direct, it inevitably becomes more incomplete or erroneous, entering the third sphere, that of *speculation*.

The sphere of *speculation* begins within the sphere of mediation, and extends outwards, indexical linkages growing thinner and thinner, until it merges with the last sphere, that of *the unknown*, which lies beyond the known. Within the sphere of speculation, there is partial and imperfect knowledge, but, moving outward, this decreases until there is no longer any pretense for claiming something as known. In the previous section we have seen examples of speculation, present in computer imaging, computer simulation, and even in processes like fractal compression. And, of course, the speculatory filling in of details is not limited to digital media.

Knowledge and objects of knowledge can pass from one sphere to another, as they become more known or less known. For example, centuries ago, the surface of the moon, although it could be partially seen, could not be experienced any more directly than that. For a long time, it was the subject of speculation (for example, that it was made of "green cheese" or that spacemen lived on it, etc.). Telescopes provided mediated knowledge, and eventually the astronauts experienced it directly (if one ignores their spacesuits). Subatomic particles, like the top quark, have sometimes been predicted before they are found (although knowledge of them is almost always necessarily mediated, unless they are in large amounts, like electric current, which can be experienced directly). Objects of knowledge can also move in the opposite direction; a man who is experienced directly by his family dies and is experienced only through mediation —old photographs, for example— by his great-grandchildren; once those are gone, he will be little more than a name and perhaps a brief description to *their* grandchildren, and eventually there may be no record left of his existence at all.

All knowledge, then, that comes to the individual does so with a link that originates in one of these concentric spheres —direct experience, mediation, speculation— and has some connection, no matter how tenuous, back to something in the real, intersubjective (least-mediated) world. Over time, and through the development of media technology and digital technology, shifts occurred in the origins of most people's knowledge, which increased the amount of knowledge while stretching its indexical connections; and over time the balance shifted from one sphere to another.

In the days before the written word or inscribed image, people's knowledge came mainly from direct experience; they knew only a small section of the world —that which was in their immediate vicinity— and knew it well; they had to, in order to hunt or farm, and to exist. The bulk of their knowledge existed in the sphere of direct experience, with a little knowledge extending out into mediation (what others might have told them) and speculation (myths and beliefs they held). The sphere which the majority of their knowledge occupied was small, but its indexical linkage was very direct.

The rise of media —print, image, sound, and combinations of them— grew slowly at first, through time, and then faster throughout the nineteenth and especially the twentieth century, as different forms appeared. Mediation opened up whole new realms of knowledge to the general public, bringing them text, images, and sound from around the world and preserving them for future generations. People's knowledge of the world grew tremendously, and later, with more means of travel available, so did their first-hand experience of it, though not nearly as fast. Today, if one compares what one knows about the world through mediation with what one knows about it through direct, first-hand experience, it is easy to see how mediated forms of knowledge have overtaken direct ones. (Few people in the United States, for example, have been to Tibet, yet no one doubts its existence.) Not only have forms of mediated knowledge overtaken direct experience, but they in some ways have reduced direct experience, since time taken watching or listening to media is time taken away from direct experience (again, the media are directly experienced *as media*, but not as what they represent, which is the subject at issue here). According to one estimate; "The average US resident spent 1,529 hours in front of TV this year [1994] and another 1,082 hours listening to the radio."[44] Together, the two media averaged 2611 hours a year; 29.78% of a person's *total* time; and even a higher percentage if one considers only waking hours! The number seems incredibly high; but even if the average were only half of that, it would still account for a great deal of time. And these figures do not include the amount of time spent at a computer, or video games, or print media, or on the telephone, and so on. For many people, the bulk of their knowledge exists in the sphere of mediation, and some will spend more time with machines than with other people.

Digital technology extends the influence and flexibility of media —as it takes over areas dominated by analog media— and brings with it a further shift, outward into the sphere of speculation. The events produced in computer simulations —the "added detail" appearing in fractal image enhancement, interpolations of data, fantastic composited imagery, virtual reality, simulations and reconstructions— all are examples of speculation, of subjunctive realities. Speculation has always been present in people's understanding of the world; people in medieval times, for example, had many of what we would consider unfounded beliefs and superstitions about the world around them. But today, even more than the middle ages, speculation —when performed on the computer, and validated by it— has been given the authority of science, and plays a central role in many people's lives, sometimes without them knowing it. Whether openly acknowledged or quietly inherent, speculation has grown into an important epistemological tool through the use of the computer. The sphere of speculation itself represents the realm of the possible, the widest realm, but the one with the most tenuous indexical linkages and the most fragile in terms of uncertainty.

Speculation can carry forward the effects of mediation —the extending of the indexical link— or it can even call them into question and ask to what degree certain indexical linkages really exist. For example, there is Bill Kaysing, author of *We Never Went to the Moon*, a book that claims the Apollo moon landings were faked, an idea similar to plot of Peter Hyams' film *Capricorn One* (1978). Kaysing's evidence comes from a variety of sources, and there are people who agree with him, and even other books on the subject.[45] While few people believe such claims, they do make one consider how much most people rely upon extended indexical linkages and value only slightly less than direct experience.

Digital technology, and forms of mediation in general, have brought with them new ways in which a sign can be linked to a referent, extending linkages into new realms, and even questioned or displaced older forms and linkages. People remember things in mediated form, and these forms can be manipulated to increase the effect of "realism". Films like *Schindler's List* (1993) and *Ed Wood* (1994) were both made in black and white because their verisimilitude was based not on reality so much as it was on the black and white films by which their respective subjects are often known; 1940s newsreels and footage shot of the death camps in one case, and the very low budget films of Ed Wood in the other. Likewise, if asked to picture Charlie Chaplin, or for that matter, events from the turn of the century, most people will probably think of black and white imagery. The conventions of realism of older media are also appearing in new media; in some video games including the CD-ROM bestseller *Myst* (1993), there are lens flares added to views of computer-generated landscapes. The inclusion of lens flares, a photographic phenomena, in computer-generated imagery is not *realism*, but *photorealism*; this is similar to the idea behind the school of photorealist painters, who were more interested in reproducing the surface of the photograph than what it depicts. The reality depicted is that of the image, not the object depicted within it; it is the medium which is being depicted. As indexical links move into the sphere of speculation, mediated knowledge appears more and more to be what people identify as the real.

Electronic mediation has been around long enough for mediated reality to take on, for many, a status almost equal to that of direct experience. Digital media carry the status given to mediation into the realm of speculation, relying on its mathematical basis to provide what many consider to be a foundation of truth. Digital technology is only the latest step in the abstracting of reality, and it is often used to efface the reality of that abstraction which is occurring. It divides and rejoins, disintegrates and reintegrates, combines perceptual and conceptual, and presents speculation as fact. It subtly affects communication, and the media by which we see and hear the world, structuring our perception, cognition, and memories. Digital technology continues to reconfigure indexical linkages to reality and forge new ones, building new worlds and cognitive practices with which to view them, bit by bit.

NOTES

1. From Peirce's own entry for the word "Sign" in the *Dictionary of Philosophy and Psychology*, edited by James Mark Baldwin, 3 Volumes, New York: Macmillan, ©1901-1905, 2:527, and reprinted in *Peirce on Signs: Writings on Semiotic by Charles Sanders Peirce*, edited by James

Hoopes: 239-240. Chapel Hill and London: The University of North Carolina Press, ©1991.

2. Bazin, Andre, *What Is Cinema? Volume I*, essays selected and translated by Hugh Gray, Berkeley, Los Angeles, and London: University of California Press, ©1967, page 14.

3. For more on Wollheim's idea of "seeing through", see Allen, Richard, "Representation, Illusion, and the Cinema", *Cinema Journal 32*, no. 2 (Winter 1993): 21-48, and Hyslop, Alec, "Seeing through seeing-in", *The British Journal of Aesthetics* 26 (Autumn 1986): 371-379.

4. Random dot stereograms produce an illusion of depth through horizontal displacement which simulates parallax, so spatial resolution along the x-axis will play a role in determining planes of depth along the z-axis. Although as Gestalt psychologists have shown, no graphic element can be both figure and ground at the same time, certain pixels in random dot stereograms (the ones producing the three dimensional effect) come close in that they act as figure to one eye and as ground to the other eye, at the same time. Although they do not break the rules proposed by Gestalt theory, they do point out how each eye can act separately in determining which elements of a composition are figure and which are ground, and not necessarily be in agreement.

5. For a reproduction of the painting, which includes Leon Harmon's low resolution portrait of Lincoln, see De Liano, Ignacio Gomez, *Dali*, New York: Rizzoli, ©1984.

6. On image copyright law and fair use provisions, see Thompson, Kristin, "Report of the ad hoc committee of the Society for Cinema Studies, fair usage publication of film stills", *Cinema Journal* 32 (Winter 1993): 3-20.

7. Assuming a channel depth of 8 bits for each RGBA channel, no compression of any kind, and 30 frames per second for video and 24 frames per second for both film gauges. Resolutions used in this example are the following: 720 x 486 for video; 4000 x 2160 for 35mm film; and 4000 x 5720 for 70mm IMAX film. All figures are according to a handout distributed in 1992 by Carl Rosendahl, President of Pacific Data Images.

8. Robinson, Rich, "Every PXL Tells a Story", *Los Angeles READER* (November 15, 1991): 6, 16.

9. For some explanation of fractal compression techniques, see Barry Simon, "How Lossy Compression Shrinks Image Files", *PC Magazine* (July 1993): 371-382; Carol Levin, "Images Incorporated: The Fractal Solution to Big Compression Problems", *PC Magazine* (November 24, 1992): 42; W. Wayt Gibbs, "Practical Fractal", *Scientific American* (July 1993): 107-108; and Louisa F. Anson, "Fractal Image Compression", *Byte* (October 1993): 195-202.

10. Davis, Frederic E., "My Main Squeeze: Fractal Compression", *Wired* (November 1993): 55.

11. For Bayard's picture and other examples, see pages 193-200 of Mitchell, William J. T., *The Reconfigured Eye: Visual Truth in the Post-*

photographic Era, Cambridge, Massachusetts and London, England: The MIT Press, ©1992,.

12. See plate 6. on page 19 in Naomi Rosenblum, *A World History of Photography*, New York, London, and Paris: Abbeville Press, ©1981.

13. Bazin, Andre, *What Is Cinema? Volume I*, page 15.

14. Crary, Jonathan, *Techniques of the Observer*, Cambridge, Massachusetts, and London, England: The MIT Press, ©1990, page 129.

15. Ihde, Don, *Instrumental Realism: The Interface between Philosophy of Science and Philosophy of Technology*, Bloomington and Indianapolis: Indiana University Press, ©1991, pages 87-88.

16. Ibid., page 85.

17. Ibid., pages 107-108. The Ackerman quotes come from Robert Ackerman, *Data, Instruments, and Theory*, Princeton, New Jersey: Princeton University Press, ©1985, pages 31 and 9, respectively.

18. The video camera has likewise been given the status of reliable witness, in the courtroom as well as in psychotherapy sessions, where it has been shown to help mental patients whose responses and behavior are recorded and analyzed. As early as 1977 video's potential was recognized, as in the article "Videocassettes used as diagnostic tools for the mentally ill" by Dr. Arthur Parkinson in *Millimeter* (June 1977): 28. Not only were patients recorded, but a computer was then used to analyze their behavior; "it permits us to compare facial expression and physical attitude with specific responses."

19. According Ihde, an early use of false coloring was for the study of transparent tissue, in order to make cells more visible. See page 88 of Ihde, Don, *Instrumental Realism: The Interface between Philosophy of Science and Philosophy of Technology*, Bloomington and Indianapolis: Indiana University Press, ©1991.

20. Malin, David F., "A Universe of Color", *Scientific American*, August 1993, pages 73 and 75. The astronomers mentioned at the end of Malin's quote are a good example of a quantizing perception.

21. Breuker, Horst, and Hans Dreverman, Christoph Grab, Alphonse A. Rademakers, and Howard Stone, "Tracking and Imaging Elementary Particles", *Scientific American* (August, 1991): 58-63. The illustration on page 61 is especially interesting, as it shows images from three different types of image technologies side by side.

22. Ibid., page 63.

23. On the tracking of birds, see Warren, Brad, "The Winged Wired", *Wired* (August 1994): 36. On the tracking of human beings, see Baerson, Kevin M., "NOAA aids in global search and rescue", *Federal Computer Week* 6, no. 3 (February 3, 1992): 8(2). According to Baerson's article, some 2,262 individuals owe their lives to the Cospas/Sarsat, a satellite tracking system used for locating downed aircraft and sinking boats and ships.

24. Culhane, Garrett, "Mission to Planet Earth", *Wired* (December 1993): 94.

25. Of course there is not way to adequately represent objects of more than three dimensions in visual form; only the "shadow" of a hypercube can be represented in three dimensions. In order to represent more than three dimensions, some set of visual conventions must be applied, placing such images further into the symbolic realm than the iconic. For a discussion of some of these conventions, see Lloyd A. Treinish, "Inside Multidimensional Data", *Byte* (April 1993): 132-135.
26. Novitski, B. J., "Visiting lost cities; using 3D CAD and rendering tools, archaeologist and architects are making famous buildings and cities accessible to a wide audience of "travelers"", *Computer Graphics World* 16, no. 1 (January 1993): 48. On the acoustic recreation of architectural spaces, see Frauenfelder, Mark, "Listening to a Blueprint", *Wired* (February 1995): 47.
27. Bylinsky, Gene, "The Payoff from 3-D Computing", *Fortune* (Autumn, 1993): 40.
28. From "Take the Guesswork out of Modeling" in *Exchange* (Spring 1994): 12. *Exchange* is a newsletter produced by Viewpoint Datalabs, and is available from the company.
29. From *SIGGraph Video Review, Issue #49: Visualization in Scientific Computing*, "Simulated Treatment of an Ocular Tumor", Contact: Wayne Lytle, Cornell National Supercomputer Facility, B49, Caldwell Hall, Garden Avenue, Ithaca, New York, 14853.
30. Bylinsky, Gene, "The Payoff from 3-D Computing", *Fortune* (Autumn, 1993): 34 and 40.
31. Joscelyne, Andrew, "A Totally Unreal Car", *Wired*, August 1994, page 35.
32. Marcotte, Paul, "Animated Evidence: Delta 191 crash recreated through computer simulations at trial", *ABA Journal* (December 1989): 52-56. This article was the *Journal*'s cover story, and it described a number of other cases in which computer simulations of events became the deciding factor in the outcome of the case, and also tells of a company, Graphic Evidence of L. A., which specializes in producing computer simulated evidence for the courtroom.
33. Clery, Daniel, "Black box flight recorder 'film' of crashes", *New Scientist* (October 7, 1992): 19.
34. Rugoff, Ralph, "Crime Storyboard: Welcome to the post-rational world", LA Weekly (December 2-December 8, 1994): 39. For other examples of the computer in the courtroom, see David Sims, "Virtual Evidence on Trial", *IEEE Computer Graphics and Applications* (March 1993): 11-13. On Legal Video Services, a company producing products for video and computer use in the courtroom, see Gruber, Jordan, "Persuasion on the Fly", *Wired* (April 1995): 48.
35. Weber, Jack, "Visualization: seeing is believing; grasp and analyze the meaning of your data by displaying it graphically", *Byte* 18, no. 4 (April 1993): 128.

36. For more examples of the dangers to public safety caused by software glitches, see Littlewood, Bev, and Lorenzo Strigini, "The Risks of Software", *Scientific American* (November 1992): 62; and W. Wayt Gibbs, "Software's Chronic Crisis", *Scientific American* (September 1994): 86.
37. Roszak, Theodore, *The Cult of Information: A neo-Luddite treatise on high tech, artificial intelligence, and the true art of thinking*, 2nd edition, Berkeley, California: University of California Press, ©1994. Computer simulation-based decision making has being going on for some time; for a look at what it was like over twenty years ago, see United States General Accounting Office, "Advantages and limitations of computer simulation in decision making", Report to the Congress, by the Comptroller General of the United States, Washington D. C.: Department of Defense, ©1973.
38. From Simson L. Garfinkel's interview with Peter G. Neumann. See Simson L. Garfinkel, "The Dean of Disaster", *Wired* (December 1993): 46.
39. See "Faulting the numbers", *Science News* (August 24, 1991): 127.
40. Littlewood, Bev, and Lorenzo Strigini, "The Risks of Software", *Scientific American* (November 1992): 62.
41. Ibid., page 63.
42. Gibbs, W. Wayt, "Software's Chronic Crisis", *Scientific American* (September 1994): 86-87.
43. Baker, Peter, "Flawed chip brings fame", *The Milwaukee Journal* (December 17, 1994): front page.
44. Steinberg, Steve G., "Raw Data: Using Media", *Wired* (December 1994): 62.
45. On Kaysing's book, see Roger Van Bakel, "The Wrong Stuff", *Wired* (September 1994): 108.

EPILOGUE

> It's paradoxical that where people are the most closely crowded, in the big coastal cities in the East and West, the loneliness is the greatest. Back where people were so spread out in western Oregon and Idaho and Montana and the Dakotas you'd think the loneliness would have been greater, but we didn't see it so much. . . . in Montana and Idaho the physical distances are big but the psychic distances between people are small, and here it's reversed. . . . There's this primary America of freeways and jet flights and TV and movie spectaculars. And people caught up in this primary America seem to go through huge portions of their lives without much consciousness of what's immediately around them. The media have convinced them that what's around them is unimportant. And that's why they're lonely. You see it in their faces. First the little flicker of searching and then when they look at you, you're just kind of an object. You don't count. You're not what they're looking for. You're not on TV. . . .
>
> Technology is blamed for a lot of this loneliness, since the loneliness is certainly associated with the newer technological devices—TV, jets, freeways, and so on—but I hope it's been made plain that the real evil isn't the objects of technology but the tendency of technology to isolate people into lonely attitudes of objectivity.
>
> —Robert M. Pirsig, *Zen and the Art of Motorcycle Maintenance*[1]

When *Zen and the Art of Motorcycle Maintenance* appeared in 1974, the computer had yet to become the widespread phenomena it is today, integrated so deeply into people's lives. With the rise and spread of digital technology in the past two decades, the situation Pirsig described has, in many ways, grown worse, even as some people claim that technology will solve the problem. And as Pirsig points out, it's not the technology, but the tendency towards isolation where the real problem lies. As information, entertainment, and communication change over to electronic media, many social structures are rapidly converting over into cyberspace. Surrounding this changeover, there is often talk about how we must try to provide universal access, so that the information have-nots do not fall too far behind the information haves. Staying out of cyberspace is equated with living in poverty, and according to the media, the only "information" worth "having" is that which is on-line. But is World Wide Web access the solution to the loneliness and psychic distance Pirsig writes about?

If everyone else lives connected to a kind of virtual world created by media technology, looking at everything including themselves through such an apparatus, does a rejection of such abstracted reality imply an abstraction from social reality? Cyberspace may be a shared illusion, but at least it's shared. Some writers have suggested that television is the only universal experience that American kids have in common.[2] With the growing proliferation of programming and cable channels, even this may no longer be the case.

Should one not watch TV at all, nor indulge in popular culture, which is so often aimed at the lowest common denominator? Or would abstaining from popular culture remove the common denominator entirely? Is there any shared background that people have in common anymore, outside of mediated reality? To what degree does leaving the world of mediated social abstraction result in one being cut off from society? Without a shared background, people grow apart, not knowing how to act, react or interact; they are unable to relate to one another. They lack understanding of one another, they do not know what behavior to expect, and become unable to empathize with others. What they don't know about each other, their imaginations fill in, with images taken from film, TV, or other media, with all the stereotypes, prejudices, and wrong assumptions they promote.

In *The Virtual Community*, Howard Rheingold wrote of the debate in France and Japan as to whether or not to jump on the bandwagon and join in worldwide electronic networks on a national scale; if they didn't, they'd be left out, whereas if they did, they would face cultural upheaval and even some erasure of their cultural identity.[3] Such a problem is faced to some degree by most societies, and by each person on an individual level. In either case, certain things must be given up; and exactly what these sacrifices entail must be carefully taken into consideration before the choice is made. There will probably never come a time when every person on earth has an on-line identity, and there will always be people, perhaps even whole societies, like the Amish, who choose to remain outside of the mediated world; the question is where power will be concentrated, whose hands will it be in, and what will be done with it. How much of our lives are dependent on technology we don't fully understand? What areas of our lives are we willing to give over to that dependency? How much more dependent can we get? As the mediated world displaces or replaces the social world, it, too, will be transformed, as will the meaning of "social" itself. The balance between these two worlds will determine both of their futures, and the balance itself will be determined by the choices made by every individual.

NOTES

1. Pirsig, Robert M., *Zen and the Art of Motorcycle Maintenance*, New York: Bantam Books, ©1974, pages 321-322.
2. Kurnit, Paul D., "TV is the only universal experience American kids have in common", *Broadcasting*, Volume 122, Number 38, September 14, 1992, page 19.
3. Rheingold, Howard, *The Virtual Community: Homesteading on the Electronic Frontier*, Reading, Massachusetts: Addison-Wesley Publishing Co., ©1993, page 240.

BIBLIOGRAPHY

Ackerman, Robert, *Data, Instruments, and Theory*, Princeton, New Jersey: Princeton University Press, ©1985.

Allen, Richard, "Representation, Illusion, and the Cinema", *Cinema Journal 32*, No. 2 (Winter 1993): 21-48.

Anderson, Benedict Richard O'Gorman, *Imagined Communities: Reflections on the Origin and Spread of Nationalism*, London and New York: Verso, ©1991.

Ascher, Marcia, *Ethnomathematics: A Multicultural View of Mathematical Ideas*, Pacific Grove, California: Brooks/Cole Publishing Company, ©1991.

Aspray, William, editor, et al, *Computing Before Computers*, Ames, Iowa: Iowa State University Press, ©1990.

Augarten, Stan, *Bit by Bit: An Illustrated History of Computers*, New York: Ticknor & Fields, ©1984.

Aukstakalnis, Steve, and David Blatner, *Silicon Mirage: The art and science of virtual reality*, edited by Stephen F. Roth, Berkeley, California: Peachpit Press ©1992.

Babbage, H. P., *Babbage's Calculating Engines*, London: E. & F. N. Spoon, 1889, and Los Angeles: Tomash Publishers, 1982.

Barlow, John Perry, "The Economy of Ideas: A Framework for Rethinking Patents and Copyrights in the Digital Age", *Wired* (March 1994): 85.

Bazin, Andre, *What Is Cinema? Volume I*, essays selected and translated by Hugh Gray, Berkeley, Los Angeles, and London: University of California Press, ©1967.

Bell, T. F., *Jacquard Weaving and Designing*, London and New York: Longmans, Green, and Co., 1895.

Benedikt, Michael, editor, *Cyberspace: First Steps*, Cambridge, Massachusetts, and London, England: The MIT Press, ©1991.

Bentley, Joelle, "Photographing the miracle of life: the work of Lennart Nilsson", *Technology Review* 95, no. 8, (November/December 1992): 58(8).

Bishop, Peter, *Fifth Generation Computers: Concepts, Implementations and Uses*, Chichester, England: Ellis Horwood Limited, and New York: Halsted Press, A Division of John Wiley & Sons, ©1986.

Black, Rita Beck, "Seeing the Baby: The Impact of Ultrasound Technology", *Journal of Genetic Counseling* 1, no. 1 (1992): 46.

Booth, Rick, "The Beat of a Different Drum: The Cop MacDonald Story", *QST* (January 1993): 31.

Breuker, Horst; Hans Dreverman; Christoph Grab; Alphonse A. Rademakers; and Howard Stone, "Tracking and Imaging Elementary Particles", *Scientific American* (August 1991): 58-63.

Broudy, Eric, *The Book of Looms: A History of the Handloom from Ancient Times to the Present*, Hanover and London: Brown University Press, ©1979.

Bruton, Eric, *Clocks and Watches 1400-1900*, New York and Washington: Frederick A. Praeger, Publishers, ©1967.

Bunish, Christine, "From Movies to Special Venues: Will Computer Graphics Replace Models and Miniatures?", *ON Production and Post-production* (October 1994): 42-47.

Bunish, Christine, "Snatching the Subtleties of Movement: Motion Capture Gets Even More Sophisticated", *ON Production and Post-production* (November 1994): 42-47.

Cain, James D., "Hams Help Bail Out Southern California Town After Rains Cut Off Access, Communications", *QST* (March 1993): 80.

Cathy Stephens, "Lighting for Trouble-Free Compositing", *ON Production and Post-production* (October 1994): 30-35.

Ceruzzi, Paul E., *Reckoners: The Prehistory of the Digital Computer, From Relays to the Stored Program Concept, 1935-1945*, Westport, Connecticut and London, England: Greenwood Press, ©1983.

Cipolla, Carlo M., *Clocks and Culture 1300-1700*, New York: Walker and Company, ©1967.

Classen, Constance, *Worlds of Sense: Exploring the Senses in History and across Cultures*, London and New York: Routledge, ©1993.

Crary, Jonathan, *Techniques of the Observer*, Cambridge, Massachusetts, and London, England: The MIT Press, ©1990.

De Oliveira, Nicolas; Nicola Oxley; and Michael Petry, *Installation Art*, Washington, D.C.: Smithsonian Institution Press, ©1994.

De Sola Pool, Ithiel, *Forecasting the Telephone: A Retrospective Technology Assessment*, Norwood, New Jersey: Ablex Publishing Corporation, ©1983.

DeLanda, Manuel, *War in the Age of Intelligent Machines*, New York and Cambridge, Massachusetts: Zone Books, distributed by the MIT Press, ©1991.

Depp, Steven W. and Webster E. Howard, "Flat-Panel Displays: Recent advances in microelectronics and liquid crystals make possible video screens that can be hung on wall or worn on a wrist", *Scientific American* (March 1993): 90-97.

Dolan, Carrie, "Translating the Bible into Klingon Stirs Cosmic Debate: Some Favor a Literal Tack, Others Find That Alien: Help From the Lutherans", *The Wall Street Journal* (June 13, 1994): Front page, A4.

Drexler, K. Eric, *Engines of Creation: The Coming Era of Nanotechnology*, New York: Anchor Press, Bantam Doubleday Dell Publishing Group, Inc., ©1986.

Drexler, K. Eric, *Nanosystems: Molecular Machinery, Manufacturing, and Computation*, New York: John Wiley & Sons, ©1992.

Eames, Office of Charles and Ray, *A Computer Perspective: Background to the Computer Age*, introduction by I. Bernard Cohen, epilogue by Brian Randell. Cambridge, Massachusetts: Harvard University Press, ©1990.

Eisenstein, Sergei, *Film Form [and] The Film Sense; two complete and unabridged works*, edited and translated by Jay Leyda. New York: Meridian Books, ©1957.

Ellul, Jacques, "Preconceived Ideas About Mediated Information", in Everett M. Rogers and Francis Balle, editors, *The Media Revolution in the United States and Western Europe*, Norwood, New Jersey: Ablex Publishing Corporation, ©1985.

Ellul, Jacques, *The Technological Society*, translated by John Wilkinson, Foreword by Robert K. Merton. New York: Alfred A. Knopf, Inc., ©1967.

Erickson, Deborah, "Electronic Earful; Cochlear implants sound better all the time", *Scientific American* (November 1990): 132.

Ester, Michael, "Image Quality and Viewer Perception", *Leonardo* (Supplemental Issue 1990): 51-63.

Feynman, Richard, "Quantum Mechanical Computers", *Optic News* 11 (February 1985): 11-20.

Fielding, Raymond P., *The Technique of Special Effects Cinematography*, 4th edition, Boston, Massachusetts, and London, England: Focal Press, ©1985.

Foley, James D., *et al.*, *Computer Graphics: Principles and Practice*, Second Edition, New York and London: Addison-Wesley Publishing Company, ©1990.

Foster, George M., *Traditional Societies and Technological Change*, New York and London: Harper & Row, Publishers, ©1973.

Franke, Herbert W., *Computer Graphics - Computer Art, Second, Revised and Enlarged Edition*, Berlin, Heidelberg, New York and Tokyo: Springer-Verlag, ©1971.

Gelernter, David, *Mirrorworlds: or the Day Software Puts the Universe in a Shoebox... How it Will Happen and What it Will Mean*, New York and Oxford: Oxford University Press, ©1991.

Gibbs, W. Wayt, "Virtual Reality Check: Imaginary environments are still far from real", *Scientific American* (December 1994): 40.

Glaser, Anton, *History of Binary and Other Nondecimal Numeration*, Los Angeles, California: Tomash Publishers, ©1971.

Gombrich, E. H., *The Image and the Eye: Further Studies in the Psychology of Pictorial Representation*, Ithaca, New York: Cornell University Press, ©1982.

Goodman, Cynthia, *Digital Visions: Computers and Art*, New York: Harry N. Abrams, Inc., ©1987.

Goodman, Nelson, *Languages of Art*, Indianapolis, Indiana: Hackett Publishers, ©1976.

Guye, Samuel, and Henri Michel, *Time & Space: Measuring Instruments from the 15th to the 19th Century*, New York: Praeger Publishers, ©1970.

Hall, Doug, and Sally Jo Fifer, *Illuminating Video: An Essential Guide to Video Art*, San Francisco, California: Aperture Foundation Inc., ©1990.

Hardison, O. B., *Disappearing through the Skylight: Culture and Technology in the Twentieth Century*, New York: Viking, ©1989.

Heim, Michael, *The Metaphysics of Virtual Reality*, Oxford and New York: Oxford University Press, ©1993.

Hofstadter, Douglas R., *Metamagical Themas: Questing for the Essence of Mind and Pattern*, New York: Basic Books, ©1985.

Holzmann, Gerald J., and Björn Pehrson "The First Data Networks", *Scientific American* (January 1994): 124-129.

Hyman, Anthony, *Charles Babbage: Pioneer of the Computer*, Princeton, New Jersey: Princeton University Press, ©1982.

Ifrah, Georges, *From One to Zero: A Universal History of Numbers*, translated by Lowell Bair, Harmondsworth, Middlesex, England: Penguin Books, Ltd., ©1981, English translation ©1985.

Ihde, Don, *Instrumental Realism: The Interface between Philosophy of Science and Philosophy of Technology*, Bloomington and Indianapolis: Indiana University Press, ©1991.

Jenkins, Henry, *Textual Poachers: Television Fans & Participatory Culture*, New York: Routledge, ©1992.

Johnson, Elmer D., *History of Libraries in the Western World, Second Edition*, Metuchen, New Jersey: The Scarecrow Press Inc., ©1970.

Journal of Applied Developmental Psychology 15 (Special Issue: Effects of interactive entertainment technologies on development), no. 1, (January-March 1994): Entire Issue.

Kiesler, Sara, and Lee Sproull, "Group Decision Making and Communication Technology", *Organizational Behavior & Human Decision Processes* 52 (June 1992): 96-123.

Krantz, Michael, "Dollar a Minute: Realies, the Rise of the Experience Industry, and the Birth of the Urban Theme Park", *Wired* (May 1994): 104.

Kubelka, Peter, "Theory of Metrical Film", from, *The Avant Garde Film Reader*, edited by P. Adams Sitney. New York: New York University Press, ©1978.

Kula, Witold, *Measures and Men*, translated from the Polish by R. Szreter, Princeton, New Jersey: Princeton University Press, ©1986.

Kurnit, Paul D., "TV is the only universal experience American kids have in common", *Broadcasting* 122, no. 38 (September 14, 1992): 19.

Littlewood, Bev, and Lorenzo Strigini, "The Risks of Software", *Scientific American* (November 1992): 62.

Loveless, Richard, editor, *The Computer Revolution and the Arts*, Tampa, Florida: University of South Florida Press, ©1989.

Loy, Gareth, "Composing with Computers--a Survey of Some Compositional Formalisms and Music Programming Languages" in *Current Directions in Computer Music Research*, edited by Max V. Mathews and John R. Pierce. Cambridge, Massachusetts and London, England: The MIT Press, ©1989.

Mackintosh, Allan R., "Dr. Atanasoff's Computer", *Scientific American* (August 1988): 90-96.

Magid, Ron, "After Jurassic Park, Traditional Techniques May Become Fossils", *American Cinematographer* (December 1993): 60.

Magid, Ron, "CGI Spearheads Brave New World of Special Effects", *American Cinematographer* (December 1993): 28.

Malina, Roger F., "Digital Image—Digital Cinema: The Work of Art in the Age of Post-Mechanical Reproduction", *Leonardo* (Digital Image—Digital Cinema Supplemental Issue, 1990): 36.

Mallen, George L., "The Visualisation of Structural Complexity: Some Thoughts on the 21st Anniversary of the Displays Group", in *Computers in Art, Design and Animation*, John Lansdown, and Rae A. Earnshaw, editors. New York: Springer-Verlag, ©1989: 24.

Mandelbrot, Benoit, *Fractals: Form, Chance, and Dimension*, San Francisco: W. H. Freeman and Company, ©1977.

Marcotte, Paul, "Animated Evidence: Delta 191 crash recreated through computer simulations at trial", *ABA Journal* (December 1989): 52-56.

Marvin, Carolyn, *When Old Technologies Were New: Thinking About Electric Communications in the Late Nineteenth Century*, New York: Oxford University Press, ©1988.

Matheson, Kimberly, and Erland Hjelmquist, "The impact of computer-mediated communication on self-awareness", *Computers in Human Behavior* 4, no. 3 (1988): 221-233.

Mattick, Paul, Jr., "Mechanical Reproduction in the Age of Art", *Arts Magazine* 65 (Spring 1990): 64.

McPhail, Thomas L., *Electronic Colonialism: The Future of International Broadcasting and Communication*, Beverly Hills, California, and London, England: SAGE Publications, Inc., ©1981.

Metcalf, John Wallace, *Information Retrieval, British and American, 1876-1976*, Metuchen, New Jersey, The Scarecrow Press Inc., ©1976.

Mitchell, William J., *The Reconfigured Eye: Visual Truth in the Post-Photographic Era*, Cambridge, Massachusetts and London, England: The MIT Press, ©1992.

Moore, Linda K. S., "Money in the Third Millennium", by, *Electronic Money Flows: The Molding of a New Financial Order*, edited by Elinor Harris Solomon. Boston, Massachusetts: Kluwer Academic Publishers, ©1991.

Moreau, René, *The Computer Comes of Age: The People, the Hardware, and the Software*, translated by J. Howlett, Cambridge, Massachusetts and London, England: The MIT Press, ©1984.

Mumford, Lewis, *Technics and Civilization*, London: Routledge, ©1947.

Musser, Charles, *The Emergence of Cinema: The American Screen to 1907*, New York: Charles Scribner's Sons, ©1990.

Neff, John Hallmark, "The Exhibition in the Age of Mechanical Reproduction", *Artforum* 25 (January 1987): 86-90.

Newton, Norman T., *Design on the Land; The Development of Landscape Architecture*, Cambridge, Massachusetts: Belknap Press of Harvard University Press, ©1971.

Nilsson, Lennart, *A Child is Born: The drama of life before birth in unprecedented photographs*, photography by Lennart Nilsson, text by Axel Ingelman-Sundberg and Claes Wirsen, translated by Britt and Claes Wirsen and Annabelle MacMillan. New York: Delacorte Press (1st American edition), ©1966, 1967.

Noll, A. Michael, "The Digital Computer as a Creative Medium", *IEEE Spectrum* 4, no. 10 (October 1967): 90.

Ord-Hume, Arthur W. J. G., *Pianola: The History of the Self-Playing Piano*, London: George Allen and Unwin, ©1984.

Pimental, Ken, and Kevin Teixeira, *Virtual Reality: Through the New Looking Glass*, New York: Intel/Windcrest, ©1993.

Pizzello, Chris, "Projecting Realism With Introvision", *American Cinematographer* (December 1993): 66.

Posselt, E. A., *The Jacquard Machine Analyzed and Explained, Third Edition*, Philadelphia: E. A. Posselt, Publisher, and London: Sampson Low, Marston & Co., Limited, 1893.

Postman, Neil, *Technopoly: The Surrender of Culture to Technology*, New York: Knopf, ©1992.

Regan, E. C., and K. R. Price, "The frequency of occurrence and severity of side-effects of immersion virtual reality", *Aviation, Space, & Environmental Medicine* 65, no. 6 (June 1994): 527-530.

Reps, John, *The Making of Urban America*, Princeton, New Jersey: Princeton University Press, ©1965.

Rheingold, Howard, *The Virtual Community: Homesteading on the Electronic Frontier*, Reading, Massachusetts: Addison-Wesley Publishing Co., and New York: HarperCollins Publishers, ©1993.

Rheingold, Howard, *Virtual Reality*, New York: Summit Books, Simon & Schuster, Inc., ©1991.

Ritchin, Fred, *In Our Own Image: The Coming Revolution in Photography*, New York: Aperture, ©1990.

Roblin, Jean, *The Reading Fingers: Life of Louis Braille 1809-1852*, translated from the French by Ruth G. Mandalian. New York: American Foundation for the Blind, ©1955.

Rosenblum, Naomi, *A World History of Photography*, New York, London, and Paris: Abbeville Press, ©1981.

Roszak, Theodore, *The Cult of Information: A Neo-Luddite Treatise on High-Tech, Artificial Intelligence, and the True Art of Thinking*, Berkeley and Los Angeles, California: University of California Press, ©1994.

Rothenberg, Jeff, "Ensuring the Longevity of Digital Documents", *Scientific American* (January 1995): 42.

Rotman, Brian, *Signifying Nothing: The Semiotics of Zero*, New York: St. Martin's Press, ©1987.

Russett, Robert, and Starr, Cecile, *Experimental Animation: Origins of a New Art*, New York: Van Nostrand Reinhold Co., ©1976.

Schwartz, Lillian, with Laurens R. Schwartz, *The Computer Artist's Handbook: Concepts, Techniques, and Applications*, New York and London: W. W. Norton and Company, ©1992.

Shannon, Claude E., "A Symbolic Analysis of Relay and Switching Circuits", *Transactions of the AIEE* 57 (December 1938): 713-723.

Shannon, Claude E., "The Mathematical Theory of Information", *Bell Systems Technical Journal* (July 1948): 380.

Shay, Don, and Jody Duncan, *The Making of Jurassic Park*, New York: Ballantine Books, ©1993.

Shurkin, Joel, *Engines of the Mind: A History of the Computer*, New York: Norton, ©1984.

Sibley, E. H., "Alphabets & Languages", *Communications of the ACM* (May 1990): 489-490.

Simon, Barry, "How Lossy Compression Shrinks Image Files", *PC Magazine* (July 1993): 371-382.

Singh, Joseph Amrito Lal, and Zingg, Robert M., *Wolf-children and Feral Man*, Hamden, Connecticut: Archon Books, ©1966

Skjellum, A., "Making Languages English Independent", *Dr. Dobbs Journal* (September 1984): 8-10.

Smilowitz, Michael; Chad D. Compton; and Lyle Flint, "The Effects of Computer-Mediated Communication on an Individual's Judgment: A study based on the effects of Asch's social influence experiment", *Computers in Human Behavior* 4 (1988): 311-321.

Smith, Ben, "Around the World in Text Displays", *Byte* (May 1990): 268.

Smith, Thomas G., *Industrial Light and Magic: The Art of Special Effects*, New York: Ballantine Books, ©1986.

Solomon, Elinor Harris, *Electronic Money Flows: The Molding of a New Financial Order*, edited by Elinor Harris Solomon. Boston, Massachusetts: Kluwer Academic Publishers, ©1991.

Stilgoe, John R., *Common Landscape of America, 1580 to 1845*, New Haven, Connecticut: Yale University Press, ©1982.

Stipp, David, "Some computers manage to fool people at the game of imitating human beings", *The Wall Street Journal* (November 11, 1991): B58(W), B4C(E).

Stix, Gary, "Encoding the "Neatness" of Ones and Zeroes", *Scientific American* (September, 1991): 54.

Stix, Gary, "Gene readers; microelectronics has begun to merge with biotechnology", *Scientific American* (January 1994): 149-150.

Stix, Gary, "Micron Machinations", *Scientific American* (November 1992): 107.

Strain, Ellen, "Mastering New Worlds: Tourists in Virtual Reality", a paper given at the *Visible Evidence II Conference*, 1994, at the University of Southern California, August 18-21.

Tanizaki, Junichiro, *In Praise of Shadows*, Translated by Thomas J. Harper and Edward G. Seidensticker, New Haven, Connecticut: Leete's Island Books, ©1977.

Tayli, Murat, and Abdulla I. Al-Salamah, "Building Bilingual Microcomputer Systems", *Communications of the ACM* (May 1990): 496-497.

Thalmann, Nadia Magnenat and Daniel, *Synthetic Actors in Computer-Generated 3-D Films*, Berlin, New York, London: Springer-Verlag, ©1990.

Thomas, Frank, and Johnston, Ollie, *Disney Animation: The Illusion of Life*, New York: Abbeville Press, ©1981.

Thompson, Kristin, "Report of the ad hoc committee of the Society for Cinema Studies, fair usage publication of film stills", *Cinema Journal*, 32 (Winter 1993): 3-20.

Thomson, F. Paul, *Money in the Computer Age*, Oxford and New York: Pergamon Press, ©1968.

Tunstall, Jeremy, "The American Role in Worldwide Mass Communication", in *Mass Media Policies in Changing Cultures*, edited by George Gerbner. New York and London: John Wiley & Sons, ©1977.

Turner, Anthony John, *Of Time and Measurement: Studies in the History of Horology and Fine Technology*, Brookfield, Vermont: Variorum, ©1993.

Valacich, Joseph S.; Alan R. Dennis; and J. F. Nunamaker, "Group Size and Anonymity Effects on Computer-Mediated Idea Generation", *Small Group Research* 23 (February 1992): 49-73.

Weber, Jack, "Visualization: seeing is believing; grasp and analyze the meaning of your data by displaying it graphically", *Byte* 18, no. 4 (April 1993): 128.

Wickramasinghe, H. Kumar, "Scanned-probe Microscopes", *Scientific American* 261 (October 1989): 101.

Wilder, Clinton, "Virtual reality seeks practicality: firm's drive toward real-world applications shows promise of potentially big market", *Computerworld* 26, no. 17 (April 27, 1992): 26.

Williams, Michael R., *A History of Computing Technology*, Englewood Cliffs, New Jersey: Prentice-Hall, Inc., ©1985.

Witold Kula, *Measures and Men*, translated from the Polish by R. Szreter, Princeton, New Jersey: Princeton University Press, ©1986.

Youngblood, Gene, *Expanded Cinema*, New York: Dutton, ©1970.

Zupko, Ronald Edward, *Revolution in Measurement: Western European Weights and Measures Since the Age of Science*, Philadelphia, Pennsylvania: American Philosophical Society, ©1990.

INDEX

Note: Some general entries appearing all throughout the book, such as "digital technology" or "analog technology", are not listed in the Index.

About the Author

Mark. J. P. Wolf teaches in the Communication Department at Concordia University Wisconsin. He has a Bachelor of Arts, Master of Arts, and Ph. D. from the School of Cinema/Television at the University of Southern California in Los Angeles. He has essays published in anthologies and has written for *Film Quarterly*, *The Velvet Light Trap*, and *The Spectator*.